John Shertzer Hittell

Mining in the Pacific States of North America

John Shertzer Hittell

Mining in the Pacific States of North America

ISBN/EAN: 9783337186111

Printed in Europe, USA, Canada, Australia, Japan

Cover: Foto ©berggeist007 / pixelio.de

More available books at **www.hansebooks.com**

MINING

IN THE

PACIFIC STATES

OF

NORTH AMERICA.

By JOHN S. HITTELL.

SAN FRANCISCO:
H. H. BANCROFT AND COMPANY.
1861.

PREFACE.

This book is intended for the general reader, not for the specialist. I do not undertake to teach the quartz miner or the hydraulic miner anything new about his special branch of business, but I may tell something about the occupation of each to the other. My object has been to collect, arrange in a lucid manner, condense and set forth in a clear style, all the attainable information about the main points of our mining and mineral resources, so that the reader, who has never been in the mines, can easily form an intelligible idea of the chief industry of this coast. Specialists might desire more elaborate treatment of their various branches, but a thorough technical handling of geology, chemistry, metallurgy, mechanics, law, history and trade, in so far as they are connected with mining in California, would have required an encyclopedia. Besides, accurate information on many points is not to be had, or can be obtained only by great expense.

I am indebted for useful hints to Messrs. C. Heusch, C. S. Capp, Dr. J. A. Veatch, C. Trinius and C. Wennerhold.

In the second chapter, I have expressed the opinion that California never can produce her own coal, and this has been, until very recently, the opinion of all the geologists in the State; but since that chapter was written, the coal mines on the slope of Mount Diablo have been opened, and Professor Whitney, State Geologist, thinks they will probably furnish good coal. I am not convinced, but dare not say much against such authority.

The fifth chapter, in which Washoe is spoken of as part of Utah, was in type before we had news of the organization of Nevada Territory.

JOHN S. HITTELL.

SAN FRANCISCO, April 1st, 1861.

INDEX TO CHAPTERS.

CHAP.		PAGE.
I.	HISTORY OF MINING IN THE PACIFIC STATES	9
II.	MINERALOGY OF GOLD	41
III.	CHEMISTRY OF GOLD	49
IV.	GEOLOGY OF GOLD	54
V.	THE MINING DISTRICTS	68
VI.	PROSPECTING	115
VII.	ASSAYING	121
VIII.	MODES OF PLACER MINING	127
IX.	PROCESSES OF GOLD-QUARTZ MINING	154
X.	PROCESSES OF SILVER MINING	168
XI.	THE LAWS OF MINING IN CALIFORNIA	176
XII.	MISCELLANY	204
	APPENDIX	214

SEE PARTICULAR INDEX ON NEXT PAGE.

PARTICULAR INDEX.

CHAPTER I. HISTORY OF MINING IN THE PACIFIC STATES, page 9.

	PAGE.		PAGE.
§ 1.	Discovery of CalifornianGold, 9	§ 20.	Mining Excitements......... 22
§ 2.	Drake's Report............... 9	§ 21.	The Greenwood Rush...... 22
§ 3.	Spanish Reports............. 10	§ 22.	Gold Lake................. 23
§ 4.	Forbes and Maufras......... 10	§ 23.	Gold Bluff................ 23
§ 5.	Dana........................ 10	§ 24.	Second Gold Lake 25
§ 6.	Larkin 11	§ 25.	Australia.................. 26
§ 7.	Marshall, the true discoverer 11	§ 26.	Peru....................... 26
§ 8.	Another version............. 13	§ 27.	Small Rushes.............. 27
§ 9.	Mining becomes a business.. 14	§ 28.	Kern River................. 27
§ 10.	New Placers found 15	§ 29.	Sacramento and Oakland.... 29
§ 11.	Newspaper Reports.......... 16	§ 30.	Fraser River............... 29
§ 12.	Rush to the Mines........... 17	§ 31.	The Washoe Fever.......... 35
§ 13.	Excitement in the States.... 18	§ 32.	Mining Inventions of California............................ 36
§ 14.	Excitement in Europe....... 14		
§ 15.	Discovery of New Districts.. 20	§ 33.	Exhaustion of the Mines.... 39
§ 16.	Number of Miners........... 20	§ 34.	Various changes since '49.... 38
§ 17.	Wages of Miners............. 21	§ 35.	New Almaden.............. 38
§ 18.	Character of Claims......... 21	§ 36.	The Gold Yield............. 39
§ 19.	Mining Implements.......... 22		

CHAPTER II. MINERALOGY OF GOLD.................page 41.

§ 37.	Metals obtained on the Coast 41	§ 41.	Forms of Quartz.... Gold... 43
§ 38.	No Ore of Gold.............. 42	§ 42.	Gold Dust.................. 44
§ 39.	Pure Gold 42	§ 43.	Forms of Placer Gold....... 44
§ 40.	Gold Mines 42	§ 44.	Lumps and Nuggets........ 45

CHAPTER III. CHEMISTRY OF GOLD .... page 49.

§ 45. Chemical Fineness of Gold .... 49.

CHAPTER IV. THE GEOLOGY OF GOLD .... page 54.

§ 46. The formation of Gold 54
§ 47. Quartz the Mother of Gold .. 54
§ 48. The Igneous Theory 54
§ 49. The Vapor Theory 55
§ 50. The Aqueous Theory 56
§ 51. The Country of Gold 56
§ 52. Rules of Quartz Veins 57
§ 53. Quartz Veins poorer as they descend 58

§ 54. The Great Quartz Vein of California 58
§ 55. Formation of the Placers 70
§ 56. Diluvial and Alluvial Placers 62
§ 57. Ancient and Modern Streams 63
§ 58. Blake's Classification 63
§ 59. Position of Pay Dirt 65
§ 60. Geological Character of Coast 65
§ 61. Volcanic Mountains 66
§ 62. Hot Springs 66

CHAPTER V. THE MINING DISTRICTS .... page 68.

§ 63. Orography of the Coast 68
§ 64. Metals of the Sierra Nevada 68
§ 65. General List of Districts 69
§ 66. Climates 70
§ 67. Sacramento District 71
§ 68. Plumas County 72
§ 69. Sierra County 72
§ 70. Butte County 78
§ 71. Nevada County 79
§ 72. Yuba County 81
§ 73. Placer County 84
§ 74. El Dorado County 86
§ 75. Amador County 89

§ 76. Calaveras County 90
§ 77. Tuolumne County 92
§ 78. Mariposa County 95
§ 79. Shasta District 79
§ 80. Kern River District 99
§ 81. Fraser District 100
§ 82. Rogue River 101
§ 83. Upper Columbia District 102
§ 84. Washoe 105
§ 85. Esmeralda 109
§ 86. Coso 110
§ 87. Arizona 110
§ 88. Quicksilver Districts 111

CHAPTER VI. PROSPECTING .... page 115.

§ 89. Prospecting a River Bar 115
§ 90. Prospecting in a Ravine 117
§ 91. Prospecting with a Knife 117

§ 92. Prospecting a Flat 118
§ 93. Prospecting for Quartz 118

CHAPTER VII. ASSAYING .... page 121.

§ 94. Kinds of Assays 120
§ 95. Means of Assaying 120
§ 96. Gold Assay with a Spoon ..120
§ 97. Assay of a Metallic Substance 122
§ 98. Gold Assay by Smelting 122
§ 99. Presence of Copper Pyrites ..123

§ 100. Silver Assay with Testing Tube 123
§ 101. Silver Assay by Smelting ...123
§ 102. Assaying Gold Quartz by weight 124
§ 103. Importance of fair Samples in Assaying 125

CHAPTER VIII. MODES OF PLACER MINING .... page 127.

§ 104. List of Modes..............127
§ 105. Knife Mining...............127
§ 106. Dry Diggings..............127
§ 107. Dry Washing...............127
§ 108. Panning....................129
§ 109. The Rocker................129
§ 110. The Puddling Box.........132
§ 111. The Long Tom.............133
§ 112. The Quicksilver Machine...133
§ 113. The Board Sluice...........134
§ 114. The Rock Sluice............134
§ 115. The Tail Sluice............140
§ 116. The Ground Sluice.........148
§ 117. The Sluice Tunnel.........142
§ 118. The Under-current Sluice..143
§ 119. Hydraulic Mining..........144
§ 120. Tunnel Mining..............146
§ 121. Shaft Mining..............148
§ 122. River Mining..............149
§ 123. Beach Mining..............151
§ 124. Blasting...................152
§ 125. Mining Ditches............153

CHAPTER IX. PROCESSES OF QUARTZ MINING........page 154.

§ 126. Comparison of Quartz with Placer Mining............154
§ 127. Quarrying the Rock........155
§ 128. Pulverizing the Quartz......155
§ 129. Pulverizing with a stone....155
§ 130. The Arastra................155
§ 131. The Chilean Mill...........157
§ 132. The Square Stamp..........157
§ 133. The Rotary Stamp..........158
§ 134. Horizontal Stones..........158
§ 135. Separation of Gold..........158
§ 136. Appliances for Separating..158
§ 137. The Blanket................159
§ 138. The Golden Fleece..........159
§ 139. The Sluice 159
§ 140. Amalgamation..............159
§ 141. In the Battery.............159
§ 142. Copper Slate...............160
§ 143. Amalgamating Basins......160
§ 144. General Remarks...........160
§ 145. Sulphurets and Amalgamation 166
§ 146. Quartz Mining as a business.166

CHAPTER X. PROCESSES OF SILVER MINING......... page 168.

§ 147. Comparison of Gold and Silver Mining................168
§ 148. Silver Ores of Pacific Coast.168
§ 149. The Reduction of Silver Ores 168
§ 150. Silver Smelting............170
§ 151. Salt Solution Process......170
§ 152. Barrel Amalgamation......171
§ 153. The Patio Process..........172
§ 154. The Eliquation Process 173
§ 155. New Processes.............173

CHAPTER XI. THE LAWS OF MINING IN CALIFORNIA, page 176.

§ 156. Public Mining Land open to all........................176
§ 157. Mineral Land owned in fee.176
§ 158. Foreign Miners............179
§ 159. Authority of Miners' Regulations..................180
§ 160. Mode of taking up Claims..181
§ 161. Incorporated Mining Companies....................182
§ 162. Decisions of Supreme Court on Mining................183
§ 163. Mining Law in Nevada.....184
§ 164. Mining on Pre-emption Claims...................184

PARTICULAR INDEX.

§ 165. Limits of Claims downwards 184
§ 166. Fluming and tailing Claims. 186
§ 167. Conveyance of Mining Claims.....................187
§ 168. Water subject to claim.....187
§ 169. Older claim has preference.189
§ 170. Abandonment of Water....189
§ 171. Claims entitled to Water...189
§ 172. Claim of water for a Ditch..189
§ 173. Miners' Regulations........189
§ 174. Regulations of Columbia District...................192
§ 175. Regulations of Pilot Hill....195
§ 176. Regulations of Mush Flat...195
§ 177. Regulations of New Kanaka Camp........................197
§ 178. Quartz Regulations of Tuolumne County............198
§ 179. Silver Regulations of Virginia District.............199
§ 180. Silver Regulations of Genoa District 200

CHAPTER XII. MISCELLANY.......................page 204.

§ 181. Value of Gold according to fineness..................204
§ 182. Gold Dust trade in California..... 204
§ 183. How Gold is carried........205
§ 184. Private Coins of California..208
§ 185. Gold Export of California...208
§ 186. Cheats in Mining...........208
§ 187. How the Miners live........211
§ 188. Cost of living..............212
§ 189. Mineral Lands should be sold.....................212

APPENDIX.....................................page 214

CHAPTER I.

HISTORICAL SKETCH.

Discovery of Californian Gold. § 1. The existence of rich and extensive gold mines in California was discovered by James W. Marshall, an American citizen and a native of New Jersey, on the nineteenth of January, 1848. Gold had, previous to that time, been found, but in places where the mines were not extensive; their production was scarcely known to commerce, and their working, after long years, led to no important results. Marshall's discovery speedily and directly exercised an influence that was felt throughout the world, and gave a new life to trade and industry in Europe and America.

Drake's Report. § 2. The first published report of gold in California, is in Hakluyt's account of Sir Francis Drake's visit to this coast in 1579. That voyager entered a bay, about latitude thirty-eight degrees, supposed to be the one now called "Drake's Bay," twenty miles north-westward from the mouth of San Francisco Bay. If not the former, it certainly was the latter bay. The historian of the voyage says: "There is no part of the earth here to be taken up wherein there is not a reasonable quantity of gold or silver." There is no statement that any of Drake's men penetrated into the interior, or made any search for these metals, or obtained any specimens of them; and since neither gold nor silver is found in the loose earth at either Drake's Bay or San Francisco Bay, we are justified in presuming the statement to be an impudent lie, written for the purposes of making the voyage appear important, giving interest to the narrative, and imposing on the ignorant and credulous.

Spanish Reports. § 3. The Spaniards and Mexicans who visited the coast at various times, by land and sea, and who were familiar with the indications of the precious metals and knew how to search for them, undoubtedly found gold at various places, particularly near the Colorado river; but they found no placers rich enough to pay for the labor of working. The impression went abroad, however, that the country had great mineral wealth, and continued to prevail until the American conquest. It was only a vague rumor, and was published in several books, but it could not command the confidence of severe criticism.

Forbes and Maufras. § 4. It is reported that silver was discovered at Alizal, in Monterey county, as early as 1802, and gold was found at San Isidro, in San Diego county, in 1828, (*Maufras*, vol. 1, p. 335); but the former place never yielded any silver worthy of note, and the latter had not been heard of in 1835, by Alexander Forbes, the historian of California, who wrote: "No minerals of particular importance have yet been found in Upper California, nor any ores of metals." (P. 173.) In another place, (p. 143) referring to Hijar's migration to California in 1833, he says: "There were goldsmiths [in the party] proceeding to a country where no gold existed." The first mine to produce any noteworthy amount of precious metal was the gold placer in the Canon of San Francisquito, on the ranch of the same name, forty-five miles north-north-westward from Los Angeles. This placer was discovered about the year 1838, (*Maufras*, vol. 1, p. 137) and in 1842 the chief miner there was a Frenchman named Barée. This placer was wrought continuously from 1838 till 1848, when it was deserted for the richer diggings in the Sacramento basin. The total yield in ten years was probably not over $60,000, a yearly average of $6,000.

Dana. § 5. In 1842, James D. Dana, the geologist and mineralogist of Wilkes' Exploring Expedition, visited California, and traveled from the northern boundary through the Sacramento basin to the Bay of San Francisco, and soon after his return to the Eastern States in 1842 or 1843, he published a work on mineralogy, in which he asserted the existence of gold in California. I have not been able to find a copy of the first edition of his book, but a newspaper which has fallen into my hands gives the following quotations, presumed to be correct.

Speaking of places where gold has been found, he mentioned—"California, between the Sierra Nevada and Sacramento and San Joaquin rivers."—(p. 251.) On page 252 he says: "The gold rocks and veins of quartz were observed by the author in 1842, near the Umpqua river, in Southern Oregon, and pebbles from similar rocks were met with along the shores of the Sacramento, in California, and the resemblance to other gold districts was remarked, but there was no opportunity of exploring the country at the time." Mr. Dana unquestionably discovered the existence of gold in California, either by direct vision or by inference, but it was a mere nominal discovery, creditable in a scientific point of view, but of no practical use. He did not find diggings that would pay, nor did his announcement set anybody to work to hunt for such diggings. His merit, in so far as California is concerned, may be compared to that of Murchison's similar discovery of auriferous rock, or rock indicating auriferous wealth, in Australia. It did no good, and nobody paid any attention to it, until the paying diggings were found by Hargraves, many years later. As Hargraves is the hero of the Australian, so is Marshall of the Californian gold discovery.

Larkin. § 6. Before giving the account of his discovery, however, I will quote the following passage from a letter, written on the fourth of May, 1846, by Thomas O. Larkin, then U. S. Consul at Monterey, California, to James Buchanan, Secretary of State under President Polk:

"There is said to be black lead in the country at San Fernando, near San Pedro, [now Los Angeles county]. By washing the sand in a plate, any person can obtain from $1 to $5 per day of gold that brings $17 per ounce in Boston; the gold has been gathered for two or three years, though but few have the patience to look for it. On the southeast end of the Island of Catalina, there is a silver mine from which silver has been extracted. There is no doubt but that gold, silver, quicksilver, copper, lead, sulphur and coal mines, are to be found all over California, and it is equally doubtful whether, under their present owners, they will ever be worked."

Marshall the true Discoverer. § 7. James W. Marshall, in a letter dated January 28th, 1856, and addressed to Charles E. Pickett, gave the following account of the gold discovery:

"Towards the end of August, 1847, Capt. Sutter and I formed a copartnership to build and run a saw-mill upon a site selected by myself (since known as Coloma). We employed P. L. Weimer and family to remove from the Fort [Sutter's Fort] to the mill site, to cook and labor for us. Nearly the first work done was the building of a double log cabin, about half a mile from the mill site. We commenced the mill about Christmas. Some of the mill hands wanted a cabin near the mill. This was built, and I went to the Fort to superintend the construction of the mill irons, leaving orders to cut a narrow ditch where the race was to be made. Upon my return, in January, 1848, I found the ditch cut as directed, and those who were working on the same were doing so at a great disadvantage, expending their labor upon the head of the race instead of the foot.

"I immediately changed the course of things, and upon the nineteenth of the same month of January, discovered the gold near the lower end of the race, about two hundred yards below the mill. William Scott was the second man to see the metal. He was at work at a carpenter's bench near the mill. I showed the gold to him. Alexander Stephens, James Brown, Henry Bigler and William Johnston were likewise working in front of the mill, framing the upper story. They were called up next, and, of course, saw the precious metal. P. L. Weimer and Charles Bennett were at the old double log cabin (where Hastings & Co. afterwards kept a store) and in my opinion, at least half a mile distant.

"In the meantime we put in some wheat and peas, nearly five acres, across the river. In February the Captain [Captain Sutter] came to the mountains for the first time. Then we consummated a treaty with the Indians, which had been previously negotiated. The tenor of this was that we were to pay them $200 yearly in goods, at Yerba Buena prices, for the joint possession and occupation of the land with them; they agreeing not to kill our stock, viz.: horses, cattle, hogs or sheep, nor burn the grass within the limits fixed by the treaty. At the same time, Capt. Sutter, myself and Isaac Humphrey entered into a copartnership to dig gold. A short time afterwards P. L. Weimer moved away from the mill, and was away two or three months, when he returned. With all the events that subsequently occurred, you and the public are well informed."

The above is the most precise, and is generally considered to be the most correct account of the gold discovery.

Another Version. § 8. Other versions of the story have been published, however, and the following, from an article published in the Coloma *Argus*, in the latter part of the year 1855, is one of them. The statement was evidently derived from Weimer, who lives at Coloma:

"That James W. Marshall picked up the first piece of gold, is beyond doubt. Peter L. Wimmer, [Weimer] who resides in this place, states positively that Mr. Marshall picked up the gold in his presence; they both saw it, and each spoke at the same time—'What's that yellow stuff?' Marshall being a step in advance, picked it up. This first piece of gold is now in the possession of Mrs. Wimmer, and weighs six pennyweights, eleven grains. The piece was given to her by Marshall himself. * * * The dam was finished early in January, the frame for the mill also erected, and the flume and bulkhead completed. It was at this time that Marshall and Wimmer adopted the plan of raising the gate during the night to wash out sand from the mill-race, closing it during the day, when work would be continued with shovels, etc. Early in February, the exact day is not remembered, in the morning, after shutting off the water, Marshall and Wimmer walked down the race together to see what the water had accomplished during the night. Having gone about twenty yards below the mill, they both saw the piece of gold before mentioned, and Marshall picked it up. After an examination, the gold was taken to the cabin of Wimmer, and Mrs. W. instructed to boil it in saleratus water; but she being engaged in making soap, pitched the piece into the soap kettle, where it was boiled all day and all night. The following morning the strange piece of stuff was fished out of the soap, all the brighter for the boiling it had received. Discussion now commenced, and all expressed the opinion that perhaps the yellow substance might be gold. Little was said on the subject; but every one each morning searched in the race for more, and every day found several small scales. The Indians also picked up many small thin pieces, and carried them always to Mrs. Wimmer. About three weeks after the first piece was obtained, Marshall took the fine gold, amounting to between two and three ounces, and went below to have the strange metal tested. On his return, he informed Wimmer that the stuff was gold. All hands now began to search for 'the root of all evil.' Shortly after, Captain Sutter came to Coloma, when he and Marshall assembled the Indians, and bought of them a large tract of country about Coloma, in exchange for a lot of beads and a few cotton handkerchiefs. They, under color of this Indian title, required one-third of all the gold dug on their domain, and collected at this rate until the fall of 1848, when a mining party from Oregon declined paying 'tithes,' as they called it.

"During February, 1848, Marshall and Wimmer went down the river to Mormon Island, and there found scales of gold on the rocks. Some weeks later they sent a Mr. Henderson, Sydney Willis and Mr. Fifield, Mormons, down there to dig, telling them that that place was better than Coloma. These were the first miners at Mormon Island."

Mining becomes a Business. § 9. Marshall was a man of an active, enthusiastic mind, and he at once attached great importance to his discovery. His ideas, however, were vague; he knew nothing about gold mining; he did not know how to take advantage of what he had found. Only an experienced gold miner could understand the importance of the discovery, and make it of practical value to all the world. That gold miner, fortunately, was near at hand; his name was Isaac Humphrey. He was residing in the town of San Francisco, in the month of February, when a Mr. Bennett, one of the party employed at Marshall's mill, went down to that place with some of the dust to have it tested; for it was still a matter of doubt whether this yellow metal really was gold. Bennett told his errand to a friend whom he met in San Francisco, and this friend introduced him to Humphrey, who had been a gold miner in Georgia, and was therefore competent to pass an opinion upon the stuff. Humphrey looked at the dust, pronounced it gold at the first glance, and expressed a belief that the diggings must be rich. He made inquiries about the place where the gold was found, and subsequent inquiries about the trustworthiness of Mr. Bennett, and on the seventh of March he was at the mill. He tried to induce several of his friends in San Francisco to go with him; they all thought his expedition a foolish one, and he had to go alone. He found that there was some talk about the gold, and persons would occasionally go about looking for pieces of it; but no one was engaged in mining, and the work of the mill was going on as usual. On the eighth he went out prospecting with a pan, and satisfied himself that the country in that vicinity was rich in gold. He then made a rocker and commenced the business of washing gold, and thus began the business of mining in California. Others saw how he did it, followed his example, found that the work was profitable, and abandoned all other occupations. The news of their success spread, people flocked to the place, learned how to use the rocker, discovered new diggings, and in the course of a few months, the country had been overturned by a social and industrial revolution.

New Placers found. § 10. Mr. Humphrey had not been at work more than three or four days before a Frenchman, called Baptiste, who had been a gold miner in Mexico for many years, came to the mill, and he agreed with Humphrey that California was very rich in gold. He, too, went to work, and being an excellent prospecter, he was of great service in teaching the new-comers the principles of prospecting and mining for gold—principles not abstruse, yet not likely to suggest themselves, at first thought, to men entirely ignorant of the business. Baptiste had been employed by Capt. Sutter to saw lumber with a whip-saw, and had been at work for two years at a place, since called Weber, about ten miles eastward from Coloma. When he saw the diggings at the latter place, he at once said there were rich mines where he had been sawing, and he expressed surprise that it had never occurred to him before, so experienced in gold mining as he was; but afterwards he said it had been so ordered by Providence, that the gold might not be discovered until California should be in the hands of the Americans.

About the middle of March, P. B. Reading, an American, now a prominent and wealthy citizen of the State, then the owner of a large ranch on the western bank of the Sacramento river, near where it issues from the mountains, came to Coloma, and after looking about at the diggings, said that if similarity in the appearance of the country could be taken as a guide, there must be gold in the hills near his ranch; and he went off, declaring his intention to go back and make an examination of them. John Bidwell, another American, now a wealthy and influential citizen, then residing on his ranch on the bank of Feather river, came to Coloma about a week later, and he said there must be gold near his ranch, and he went off with expressions similar to those used by Reading. In a few weeks, news came that Reading had found diggings near Clear creek, at the head of the Sacramento valley, and was at work there with his Indians; and not long after, it was reported that Bidwell was at work with his Indians on a rich bar of Feather river, since called "Bidwell's Bar."

Newspaper Reports. § 11. Although Bennett had arrived at San Francisco in February with some of the dust, the editors of the town—for two papers were published in the place at the time—did not hear of the discovery till some weeks later. The first published notice of the gold was given in the *Californian*, (published in San Francisco) on the fifteenth of March, as follows:

"GOLD MINE FOUND.—In the newly made race-way of the saw mill recently erected by Captain Sutter, on the American Fork, gold has been found in considerable quantities. One person brought thirty dollars' worth to New Helvetia, gathered there in a short time. California, no doubt, is rich in mineral wealth; great chances here for scientific capitalists. Gold has been found in almost every part of the country."

Three days later, the *California Star*, the rival paper, gave the following account of the discovery:

"We were informed a few days since, that a very valuable silver mine was situated in the vicinity of this place, and again, that its locality was known. Mines of quicksilver are being found all over the country. Gold has been discovered in the northern Sacramento districts, about forty miles above Sutter's Fort. Rich mines of copper are said to exist north of these bays."

Although these articles were written two months after the discovery, it is evident that the editors had heard only vague rumors, and attached little importance to them. The *Star* of the twenty-fifth of March, says:

"So great is the quantity of gold taken from the new mine recently found at New Helvetia, that it has become an article of traffic in that vicinity."

None of the gold had been seen in San Francisco; but at Sutter's Fort, men had begun to buy and sell with it.

The next number of the *Star*, bearing date April 1st, 1848, contained an article several columns long, written by Dr. V. J. Fourgeaud, on the resources of California. He devoted about a column to the minerals, and in the course of his remarks, said :

"It would be utterly impossible at present to make a correct estimate of the mineral wealth of California. Popular attention has been but lately directed to it. But the discoveries that have already been made, will warrant us in the assertion that California is one of the richest mineral countries in the world. Gold, silver, quicksilver, iron, copper, lead, sulphur, saltpetre and other mines of great value have already been found. We saw a few days ago a beautiful specimen of gold from the mine newly discovered on the American Fork. From all accounts the mine is immensely rich, and already we learn the gold from it, collected at random and without any trouble, has become an article of trade at the upper settlements. This precious metal abounds in this country. We have heard of several other newly discovered mines of gold, but as these reports are not yet authenticated, we shall pass over them. However, it is well known that there is a placero of gold a few miles from the ciudad de Los Angeles, and another on the San Joaquin."

It was not until more than three months after Marshall's discovery, that the San Francisco papers stated that gold mining had become a regular and profitable business in the new placers. The *Californian* of April 26th, said :

"GOLD MINES OF THE SACRAMENTO.—From a gentleman just from the gold region, we learn that many new discoveries have very recently been made, and it is fully ascertained that a large extent of country abounds with that precious mineral. Seven men, with picks and spades, gather $1,600 worth in fifteen days. Many persons are settling on the lands with the view of holding preëmptions, but as yet every person takes the right to gather all they can, without any regard to claims. The largest piece yet found is worth six dollars."

Rush to the Mines. ¿ 12. The news spread, men came from all the settled parts of the territory, and as they came they went to work mining, and gradually they moved further and further from Coloma, and before the rainy season had commenced (in December) miners were washing rich auriferous dirt all along the western slope of the Sierra Nevada, from the Feather to the Tuolumne river, a distance of one hundred and fifty miles, and also over a space of about fifteen miles square, near the place now known as the town of Shasta, in the Coast mountains, at the head of the Sacramento valley. The whole country had been turned topsy-turvy; towns had been deserted, or left only to the women and children; fields had been left unreaped; herds of cattle went without any one to care for them. But gold mining, which had become the great interest of the country, was not neglected. The people learned rapidly and worked hard. In the latter part of 1848, adventurers began to arrive from Oregon, the Sandwich Islands and Mexico. The winter found the miners with very little preparation, but most of them were accustomed to a rough mode of life in the western wilds, and they considered their large profits an abundant compensation for their privations and hardships. The weather was so mild in December and January. that they could work almost as well as in the summer, and the rain gave them facilities for washing such as they could not have in the dry season.

Excitement in the States. § 13. In September, 1848, the first rumors of the gold discovery began to reach New York; in October they attracted attention; in November people looked with interest for new reports; in December the news gained general credence and a great excitement arose. Preparations were made for a migration to California by somebody in nearly every town in the United States. The great body of the emigrants went either across the plains with ox or mule teams, or round Cape Horn in sailing vessels. A few took passage in the steamer by way of Panama. Not less than one hundred thousand men, representing in their nativity every State in the Union, went to California that year. Of these, twenty thousand crossed the continent by way of the South Pass; and nearly all of them started from the Missouri river between Independence and St. Joseph's, in the month of May. They formed an army; in day time their trains filled up the road for miles, and at night their camp-fires glittered in every direction about the places blessed with grass and water. The excitement continued during '50, '51, '52 and '53; emigrants continued to come by land and sea, from Europe and America, and in the last named year from China also. In 1854 the migration fell off, and since that time California has received the chief accessions to her white population by the Panama steamers.

Excitement in Europe. ₴ 14. The whole world felt a beneficent influence from the great gold yield of the Sacramento basin. Labor rose in value, and industry was stimulated from St. Louis to Constantinople. The news, however, was not welcome to all classes. Many of the capitalists feared that gold would soon be so abundant as to be worthless, and European statesmen feared the power to be gained by the arrogant and turbulent democracy of the New World.

The author of a book, entitled *Notes on the Gold Districts*, published in London in 1853, thus speaks of the fears excited in Europe on the first great influx of gold from the Californian mines:

"Among the many extraordinary incidents connected with the Californian discoveries, was the alarm communicated to many classes, which was not confined to individuals, but invaded governments. The first announcement spread alarm; but as the cargoes of gold rose from $100,000 to $1,000,000, bankers and financiers began seriously to prepare for an expected crisis. In England and the United States the panic was confined to a few; but on the continent of Europe every government, rich or poor, thought it needful to make provision against the threatened evils. An immediate alteration in prices was looked for; money was to become so abundant that all ordinary commodities were to rise, but more especially the proportion between gold and silver was to be disturbed, some thinking that the latter might become the dearer metal. The governments of France, Holland and Russia, in particular, turned their attention to the monetary question, and in 1850 the government of Holland availed themselves of a law, which had not before been put in operation, to take immediate steps for selling off the gold in the Bank of Amsterdam, at what they supposed to be the then highest prices, and to stock themselves with silver. Palladium, which is likewise a superior white metal, was held more firmly, and expectations were entertained that it would become available for plating. The stock, however, is small. The silver operation was carried on concurrent with a supply of bullion to Russia for a loan; a demand for silver in Austria, and for shipment to India, and it did really produce an effect on the silver market, which many mistook for the influence of California. The particular way in which the Netherlands operations were carried on, was especially calculated to produce the greatest disturbance of prices. The ten florin pieces were sent to Paris, coined there into Napoleons, and silver five franc pieces drawn out in their place. At Paris, the premium on gold, in a few months, fell from nearly two per cent. to a discount, and at Hamburg a like fall took place. In London, the great silver market, silver rose between the autumn and the new year, from five shillings per ounce to five shillings one and five-eighths pence per ounce, and Mexican dollars from four shillings ten and one-half pence to four shillings eleven and five-eighths pence per ounce; nor did prices recover until towards the end of the year 1851, when the fall was as sudden as the rise."

Discovery of New Districts. § 15. In the spring of 1849, Reading crossed the Coast Range with a party of his Indians, and discovered rich diggings in the valley of the Trinity. In the summer of the same year, Col. Fremont discovered the mines on his ranch in the valley of the Mariposa. The next year, the diggings in the Klamath and Scott valleys were opened; and since that time no rich and extensive gold bearing district has been discovered in the State, although paying diggings have been found in a number of places. In the fall of 1854, the diggings in the valley of Kern river were found, and in 1857 those of Mono Lake, Walker's river and San Gabriel. Each of these places has a distinct mining district, where a few hundred miners are usually at work.

The gold diggings in the valley of Rogue river, Oregon, where a large number of men are now employed, were found in 1852; the mines of Carson Valley, in Utah, in the same year; those at Colville, in Washington territory, in 1853; those of the Yakima valley, on the eastern slope of the Cascade mountains in the same territory, in 1854; and those of Fraser river in 1858. The silver mines of Arizona had been known for more than a century before the Americans purchased that district. The business of mining there was commenced by American companies in 1856. The silver mines of Washoe were discovered in 1859, and those of Esmeralda and Coso, further south, in 1860.

Number of Miners. § 16. At the end of the year 1848, there were probably five thousand miners at work in the gold diggings of California, a year later forty thousand, in the latter part of 1850 fifty thousand, and now there may be a hundred thousand. Besides the one hundred thousand miners in California, there are probably five thousand in Washoe, three thousand in Oregon, fifteen hundred in British Columbia, five hundred in Washington territory, five hundred in Esmeralda and three hundred in Arizona.

Wages of the Miners. § 17. The diggings in the immediate vicinity of Coloma were not very rich as compared with those in many other places in California, but the miners made from twenty-five to thirty dollars a day each, with the rocker. Those who were the first to get to work on the rich bars of the North Fork of the American, Yuba, Feather, Stanislaus and Trinity rivers, and numerous brooks and gullies, often made from five hundred to five thousand dollars in a day, and an average of from three to five hundred dollars a day for weeks at a time, was no rarity in '48 and '49. The average amount, however, dug per day, could not safely be put at over twenty dollars at any time; though prudent, industrious miners, who were content to make that much, usually averaged much more in '48. There were many places where the gold was evenly distributed, and in these the pay was regular but seldom large. In other places the distribution was irregular, and in those the miner might dig a hundred ounces one day, and find nothing for a week after. The average amount obtained by miners in the gold mines of California, gradually decreased from twenty dollars per day in 1848 to five dollars in 1853, and three dollars in 1860.

Character of Claims. § 18. From the spring of 1848 until 1851, nearly all the mining was done in river bars, and in small ravines; the former called "wet diggings," and the latter "dry diggings," which two classes included all the mines then worked. In 1851, the miners began to work at the hills, flats and quartz veins. Gradually the river bars and ravines grew poorer, until now they do not supply more than one-sixth of the gold yield of the State, and the distinction between "wet" and "dry" diggings has been forgotten. At first all the diggings were shallow; previous to 1852, it was rare to find any one mining more than ten feet below the surface of the earth; now thousands are regularly employed in tunnels and shafts and hydraulic claims, a hundred feet or more below the original surface of the earth.

Mining Implements. § 19. In 1848 the miners began to work with the pan and the rocker; the next year they introduced the quicksilver machine, and in 1850 the tom. In this latter year also, they commenced to flume rivers and to make extensive ditches. In 1851 they invented the sluice, and the subsequent year used the hydraulic process of tearing down banks and hills. Nevada county has the credit of being the place where the tom was first used, and the sluice invented and where the ditch and hydraulic process were first introduced. The names of the individual originators of the sluice and ditch are not on record, so far as I know, but Edward E. Matteson, a native of Sterling, Connecticut, was the inventor of hydraulic mining. Nineteen-twentieths of the placer gold now dug in California is obtained with the assistance of the tom, sluice, ditch and hydraulic; and if our miners were now restricted to the instruments and processes used in 1849, they would not produce more than $5,000,000 instead of $50,000,000 per year. The Californian inventions in gold mining have not only been of immense importance to the development of the mineral wealth of this State, but must, in time, exert a strong influence elsewhere. We know that no district containing extensive placer diggings, could ever be "worked out" without the assistance of such processes as these here invented; and, therefore, many a placer heretofore deserted, because profitless under the old modes of working, will hereafter be occupied again, and probably be made to pay better than ever before. Time and experience will prove the correctness of this view.

Mining Excitements. § 20. There have been a number of excitements in California, about rich diggings reported to exist in places not previously occupied by miners. In some of these cases, the reports of auriferous wealth were entirely without foundation; in others, there was some foundation in truth for the reports, but nothing to justify the great excitement. A historical sketch of mining in California, would not be complete without a mention of these mining excitements, and the "rushes" of the miners to these new El Dorados.

The Greenwood Rush. § 21. The first rush of this kind, though but a small one as compared with those of subsequent years, took place in June, 1849, at the instigation of a mountaineer named Greenwood, who told the miners at

Coloma that he had some years before seen an abundance of gold at Truckee Lake, though at the time he did not know what the substance was. Several hundred men went to get the precious metal which Greenwood had seen, but at the end of six or seven weeks they returned tattered and destitute, without any of the gold, though they had found the place where it was supposed to be.

Gold Lake. § 22. The next rush, and a very remarkable one, was made in May, 1850, to Gold Lake, where it was asserted that gold was very abundant. This lake is a beautiful little sheet of water north-eastward from the present site of Downieville. The origin of the excitement is accounted for in different ways, but the story which has most probabilities in its favor, is that a miner named Stoddard overheard a couple of others talking about a lake where gold was to be found by the ton lying loose on the bank. Stoddard, supposing the place to be Gold Lake, commenced the next day, to tell about it as if he had seen it, and as he believed it sincerely, and was making preparations to go, others believed too, and a rush followed. Thousands of miners deserted claims where they were making from twenty dollars to forty dollars per day. to go to this lake where they hoped to get, within a few weeks, all the gold they could want. After several months of searching, the Gold Lake prospectors returned without having found anything to reward their toils and sacrifices.

Gold Bluff. § 23. In the beginning of January, 1851, a great excitement broke out in California about the gold mines on the ocean beach at Gold Bluff, in latitude 41 deg. 25 min., two hundred and seventy-five miles northward from San Francisco. The following extract from an editorial article in the *Alta California* of the ninth of January, '51, is a sample of the reports brought from those wonderful mines while the fever was at its height:

"A NEW EL DORADO.—* * * Twenty-seven miles beyond the Trinity there is a beach several miles in extent and bounded by a high bluff. The sands of this beach are mixed with gold to an extent almost beyond belief. * * The gold is mixed with the black sand in proportions of from ten cents to ten dollars the pound. * * * Mr. [John A.] Collins, the Secretary of the Pacific Mining Company, measured a patch of gold and sand, and estimates it will yield to each member of the company the snug little sum of $43,000,000, and this estimate is formed upon a calculation that the sand holds out to be one-tenth as rich as observation warrants them in supposing. Mr. Collins saw a man who had accumulated 50,000 pounds, or 50,000 tons—he did not recollect which—of the richest kind of black sand. Gen. [John] Wilson says that thousands of men cannot exhaust this gold in thousands of years."

The *Alta* of the next day (January 10th) gives a letter from M. C. Thompson to John A. Collins, in which the writer, dating "Trinity county, January 1st, 1851," says:

"Some six months since, I saw large patches, the entire length of the beach, covered with black sand, literally yellow with small, thin flakes of gold."

Mr. Thompson made affidavit of the truth of this statement before L. P. Gilky, Justice of the Peace, and C. W. Kinsey also deposed to its truth.

The same paper also contains an extract from a letter written by E. A. Rowe, on the first of January, then a Constable in Trinity county, to J. A. Collins, as follows:

"I have seen enough in one plat of sand, containing enough of gold to yield from three dollars to ten dollars per pound, to load a ship of the largest class. * * * I am now, however, confident that with the proper arrangements for amalgamating the gold, on a scale as extensive as your company is capable of doing, millions upon millions of dollars can be easily obtained every year, for more than a century to come."

Mr. Rowe made affidavit before the same Justice Gilky to the truth of his statements in this letter. The *Alta* of the succeeding day—the eleventh of January—contains the advertisements of eight vessels to sail for Red Bluff. At that time, so little was known about the distribution of gold, and the metal was so abundant, and such marvellous fortunes were being made by so many people, that nobody knew what to believe or disbelieve about such stories. A wonderful fever rose; everybody wanted to go to Gold Bluff; but, fortunately, the bubble bursted in a few days—so soon, in fact, that few persons had time to get started for the new El Dorado. Since then, Gold Bluff has been one of our biggest gold humbugs.

Yet Gold Bluff was not all humbug. There really was much gold in the sands of the beach, and many miners have done well in washing for it. The bluffs along the beach from Trinidad to Port Orford, bear resemblance to the auriferous hills of Nevada county, which are now being washed away by the hydraulic process. Along the beach a natural hydraulic washing has been in progress for thousands of years.

Second Gold Lake. § 24. In the summer of 1851, there was an excitement at Downieville about another Gold Lake, reported to be rich. The main witness to the auriferous wealth of this place was a Canadian named Deloreaux, who became the guide of a party. One of them, in an account of the expedition published in the San Andreas *Independent*, wrote thus:

"The first difficulty that presented itself to our minds was how we should transport our gold from the diggings. It was finally settled that six of the party should take all the mules, packed with gold, and deposit the same with Uncle Sam, and, returning, bring us a fresh supply of grub. You smile, dear reader, at the magnificent folly of the conceit, but be assured that it is not exaggerated in this narrative. The common opinion of those times was that the great fountains of our golden wealth were hidden amid the crags and solitudes of the snow-belt, where, guarded by wild beasts and savage men, they were only to be reached by the most fearless and iron-souled adventurers. This was the fascinating theory that gilded the distant horizon of the forty-niner, lent a charm to the roughest life, and softened his rocky couch, as nightly he wrapped his tattered blankets around his shivering frame, and lay down to dream of the golden future. The stories of the *Arabian Nights* were to be divested of their romantic extravagance somewhere in the mountains of California, and every heart was nerved to boldness to be among those to share the fruits of first discovery. The physician forgot his agonizing patient, the lawyer laid aside his brief, the minister forgot alike his God and his flock, to join in the hot pursuit of mammon.

"Returning to our party. It was amusing to hear the arguments as to 'how many dollars a mule could pack;' and the offers of each to bet on his favorite. One of the party proposed to bet that his mule—a little, short-backed institution, which was captured by one of Col. Donophan's boys at the battle of Sacramento, and then a quarter of a century old—could pack $150,000! That bet was taken by this writer. It has not yet been decided. During the discussion of these important matters, an overgrown chap from Maine suggested that we were counting our fish before they were caught. Audacious wretch! He was forthwith voted a 'muggins and disorganizer' for entertaining any doubts upon the subject. We had more faith than it requires to save sinners at a camp meeting. We followed the trapper for eight days, over the roughest and wildest part of California, when, overcome with fatigue, we came to a halt and camped. The guide, after some time, informed us that we had taken the wrong 'divide,' but that we could right ourselves by crossing a very bad cañon. Our suspicions were aroused, and some of the less sanguine did not hesitate to murmur against the alleged treachery of Deloreaux. We kept our secret, however, and putting on a willing face, readily consented to the proposition to try the cañon.

"Next day we recommenced our line of march; but after travelling all day we found no canon. That night we had a 'talk' with the trapper, who began to *if* and *and* about it, and plainly told us that we 'had better prospect the country around us.' Thereupon, some of the party gave vent to their rage, telling him that they would 'shoot the top of his head off' if he had deceived us; but he swore to the truth of his first statement, and that he would take us to the lake in two days. We laid down that night with our spirits sadly vacillating between hope and fear. In the morning our gentle, candid Deloreaux 'came up missing.'"

Most of the party, after spending weeks in the mountains, returned to the mines in a very destitute condition, but others succeeded in finding the very rich diggings about the place now known as Forest City, in Sierra county.

In August, 1853, there was some excitement about mines in Santa Cruz county on, the ranch of Don Pedro Sainsevain, and a great many persons went thither; but they found little gold, and the place was soon deserted again.

Australia. § 25. Well authenticated reports of the discovery of gold in Australia arrived in California, but they created little excitement; and although during '52 and '53 a thousand persons or more left California for the colony of Victoria, their departure was so silent and gradual as scarcely to be noticed.

Peru. § 26. In February, 1854, a great excitement arose about gold mines in Peru. The newspapers at Panama published a number of articles and letters, to the effect that gold mines of unparalled richness had been opened on the head waters of the Amazon. It was said that many miners were at work there, and that they could dig twenty-five pounds a day to the man without difficulty. Such statements, repeated time after time, purporting to come from a number of different sources, and republished in the California and Australian papers, induced about 2,000 miners to go from California and Australia to Peru, where they found that nobody knew of any such gold mines in the country as had been spoken of by the Panama papers. Some of the adventurers went over to the head waters of the Amazon, and they found the " color," but that was all.

Small Rushes. § 27. About the same time with the excitement in regard to the gold mines in Peru, there was a report of rich diggings at Santa Anita, in Los Angeles county. Many persons went there, but soon left in disgust.

In April, 1854, there was a rush to new mines on Russian river, in Sonoma county. About a thousand persons were April-fooled into a visit to that place.

In May, of the same year, there was a rumor of a gold discovery on the side of Mount Diablo, in Contra Costa county, and many went to see.

Kern River. § 28. In February and March, 1855, the Kern river fever afflicted the State. Diggings, which paid pretty well in a few spots, had been found, and it was reported that they were very extensive and as rich as those of the Sacramento Valley in 1849. The towns of Stockton and Los Angeles were especially interested in having people to go to Kern river, because they had all the trade of that place; and their newspapers accordingly were filled with glowing accounts of the richness of the mines. The following are extracts from published letters purporting to have been written at the mines:

" Wherever the miner's pick has been struck, and good *prospect* made, gold has been discovered. In many places it is exceedingly rich, vieing well with the miraculous stories of '49, and yielding quite as liberally. The process of obtaining the gold, at present, is confined wholly to *toms*, for the want of lumber. Miners are making from sixteen dollars to sixty dollars a day by the use of the tom, and much larger amounts would be obtained, but for the want of water. There is no question that the mines are very rich; and these gentleman are of the opinion that they are much richer than any ever yet discovered in California. Great numbers from Mariposa county are flocking thither; and many of the ranches in the San Joaquin valley are being deserted, the owners of the same all bound to the new diggings.

"I found on my arrival at Greenhorn Gulch, where the first discovery was made, all the miners were very busy at work on their claims, which were paying very well, ranging from eight to fifty dollars per day. Claims could not be bought unless at very extravagant prices. The day of my arrival, one party of two, at the head of the gulch, took out fifty dollars; but as they had been making from two to three ounces daily since they commenced working their claim, to them it was but a common occurrence. Another party, lower down, on the same day, made an ounce to the man; and a party at the mouth of the gulch, consisting of four men, took out one hundred dollars. The entire gulch is already taken up, and by a few men only, each one being allowed two hundred feet. New comers go about fifteen miles higher up, where they are making from eight to sixteen dollars per day. It is reported at the gulch, and firmly believed, that a party of five or six, about forty miles higher up in the mountains, have been making one hundred dollars a day to the man for the last month."

A great excitement prevailed. Five thousand persons went to Kern river. The industry of the State was seriously disturbed. Farmers abandoned their homes, miners their claims, and mechanics their shops, hoping to make fortunes in a few months at Kern river. On reaching that place, they found that the mining district was a very small one; that the rich claims had been exhausted in a few weeks, and that the best thing they could do would be to return home.

Sacramento and Oakland. § 29. In June, 1855, some gold was found on the bank of the Sacramento river a few miles below Sacramento City, and some seven hundred dollars were washed out in a couple of days by four or five men. There was a great rush to stake off claims; but the excitement died out in a few days more, in consequence of the discovery that all the gold in these diggings had come from a pair of trousers belonging to a miner who had lived there in 1849.

In January, 1856, there was a rumor of rich diggings in the plain east of Oakland, and a multitude of claims were staked off and a few holes dug, but little or no gold was found, and the claims were soon deserted.

The Fraser Fever. § 30. In 1858 came the Fraser fever, the most serious ailment (of the kind) that ever afflicted California. In March of that year, news came of the discovery of rich gold diggings on the banks of Fraser river, in British Columbia, and the reports gradually became more and more favorable, until midsummer, when the belief was general among Californians that the valley of the Fraser was as rich in gold as was that of the Sacramento in 1849. In the course of four months, 18,000 persons left the State for British Columbia, about one in twenty of the whole white population. For a time nine ocean steamers were engaged in the conveyance of passengers between San Francisco and Victoria. Between the third and tenth of July, inclusive, 3,758 persons left San Francisco by sea, and the departures would have continued in the same proportion, if it had not been that very unfavorable news arrived in San Francisco, on the tenth of July, and suddenly put an end to the fever. During the continuance of the excitement, the business of the State was turned topsy turvy. A newspaper correspondent, writing from San Francisco under date of June 21st, said:

"The rush to Fraser has made San Francisco very lively. At no time since '53, has the city presented such a busy appearance as now; nor has so much excitement pervaded the community at any time since 1849. Indeed, 1849 has come again.

"On every side, at every turn, you hear of Fraser river. Every acquaintance you meet asks whether you are going to Fraser river, or tells how he is going, or would go if he could, and enumerates your acquaintances who are going. The newspapers are full of Fraser river. The goods exposed in the streets are marked 'Fraser River,' 'Blankets for Fraser river,' 'Shirts for Fraser River,' 'Beans for Fraser river,' 'Shovels for Fraser river,' etc. Here and there you will see fixed up in front of a store, some such sign as this: 'Selling out at cost; going to Fraser river, sure as you're born.' Every few days the newsboys run through the streets shouting, "'Ere's the *Extra Alta!* Later news from Fraser river! Gold by the bushel!' The hotels are full of people on their way to the new El Dorado, and they speak of nothing but Fraser river. Occasionally you will hear a snatch of an old song adapted to the times:

> 'Oh, I'm going to Caledonia—that's the place for me;
> I'm going to Fraser river, with the washbowl on my knee.'

"We had a revival of religion here, but Fraser river knocked it cold. People care less, apparently, just now, for salvation than gold. The Coroner of this city complains that the new diggings have put an end to the suicides. Really, Fraser river is turning California upside down; it will be changed so that the old residents will scarcely know the place. Our present population is going, and strangers will take their places.

"From some of the mining towns more than a fifth of the men have already gone to Fraser. Thus, 200 have already gone from Grass Valley, which cast 900 votes last autumn, and 75 have gone from Volcano, which cast 328 votes. These, however, are extreme cases. In many places extra stages have been put on, and the steamers from Sacramento to this city are loaded every day, even when extra boats are put on. One day last week, 39 passengers left Nevada on the Sacramento stages, and 37 of them were bound for Fraser river. In Sonora, 200 persons are waiting for their turn to come down on the stages; and so it is all through the mines. Every mining camp is losing a portion of its population; some one-tenth, some even one-half. Trade to the interior is at a standstill; the traders and boarding-house keepers, gardeners and farmers, of the mining districts, are losing their debtors and customers, and as they suffer, so their creditors will suffer. The newspapers of the interior are in spasms about the excitement; they reason against it, they pray against, they ridicule it, and some of the editors would be willing to fight against it, if fighting could be of service. The depression of business in the interior is very great. Property is falling in value, and croakers are promising that grass shall next year grow in the streets of the largest mining towns. In Sacramento and Stockton there is no life, except at the stage offices, when the stages arrive, and about the steamboats when they go away. Farms and stores are offered for sale at great sacrifices; money is rising in value, and labor likewise. Mechanics and miners are demanding higher wages; carpenters have raised their prices from four and five to six dollars per day; masons. from six to seven dollars; stevedores, from five to six dollars; hodmen, from three to four dollars; firemen on the steamboats, from sixty to eighty dollars per month; and so on through all the branches of employment wherein men are hired by the day. The contractor of the Sacramento Valley Railroad has employed Chinamen, because of inability to get white men. Many of the quartz mills have stopped, and almost all will have to stop, if the present drain continues for two months more."

The same correspondent, under date of July 5th, wrote thus:

"The Fraser fever rages fiercely. Did I say 'fever?' I should have said 'mania.' Hundreds of miners come down from Sacramento and Stockton every day, and there is no abatement, but rather an increase in the furious flood. Not since '51 has the city been so full. The hotels are crowded; the dealers in hardware, clothing and provisions, and the owners of ocean steamers are making their fortunes. The river steamers and the interior stages have raised their prices, and though extraordinary boats and stages are running, they still cannot carry all who would come. In one day twenty-eight stages and wagons, loaded down with emigrants for New Caledonia, came into Stockton. All the stages running to Sacramento come in every day filled up with Fraser-fever fellows.

"Most of the emigrants, so far, have been miners, and every large mining camp has lost a considerable portion of its population.

"The Downieville *Democrat* of the 27th ult., says: 'Whole camps in the dry diggings are almost entirely deserted, and claims held a month ago at $1,000 are now offered for $100. Diggings that are paying an ounce a day are to be bought for $100, or whatever sum may be offered.'

"The North San Juan *Star*, of the same date, says: 'We heard of one claim yesterday, which two months ago was bought for $2,500, now offered for $600, and the owner goes about begging for an offer—a claim which is worth more now than it ever was before. But this is only one of the many instances we hear of, in this vicinity as well as elsewhere.'

"I could give you a multitude of such extracts, but these are enough. Empty cabins are to be seen on every side in the mining districts, and about many of them lie unclaimed cooking utensils and mining tools. The exodus of the white miners is looked upon by the Chinamen, of whom 2,000 have arrived within the last month, as particularly fortunate for them, for they go right into possession of the deserted cabins, claims and tools. Of the Indians, the Columbia correspondent of the *Alta* writes as follows:

"'I am informed as a matter of fact, that the Indians in this vicinity, having been told that the whites were going away, came into town the other day; and agreed how they would divide the brick buildings among themselves after the pale faces should leave; and other Indians went into Sonora and made a similar division there. They left the wooden buildings out, not wanting them.'

"A correspondent of the *Alta*, writing from Vallecito, says: 'Mining claims, paying from eight to ten dollars per day, are selling at from one hundred to one hundred and seventy-five dollars apiece by those infatuated with the Fraser river fever.'

"It is impossible to know how many of her inhabitants California will lose in this stampede, or what is to be its ultimate influence on her wellfare. The total number of votes to be cast at the general election in the first week in September, will probably not exceed 50,000, whereas it was about 100,000 last year. Out of thirty policemen in this city, seven have gone, and 200 of 1,050 firemen.

"Wages have risen considerably throughout the State. Sailors are getting seventy-five dollars per month, and men servants on steamboats are getting the same price. The firemen on ocean steamers get one hundred and fifty dollars.

"The price of real estate in the towns has fallen, particularly in the mining districts, where the depreciation is as much as twenty-five and fifty per cent. Sacramento, Marysville and Stockton are suffering severely. All their trade is killed. They are at the heads of steamboat navigation, whence extensive mines are supplied. But now there is no demand; there are more goods in the mines than are wanted, and shipments of many articles are coming back to this city. The ominous sign 'To Let' is alarmingly frequent in all the towns, but particularly so in Sacramento and San Francisco; for although this city has suffered less by the mania than any other place, and although many classes of business have been rendered exceedingly profitable here by it, yet other trades have been ruined. He who deals in goods needed by miners going to a new country, is in a fair way to make a fortune; he who does not, has a dull prospect before him, unless he goes with the tide, and this many are compelled to do who would be glad to remain if they could do any business in California. The boarding-house keepers, the merchants and the rum-sellers, of the interior, are compelled to shut up and follow their customers to the land of promise."

In July the emigrants began to return, and before the end of October, most of them had got back to their old homes, poorer and wiser than they started. Only about 3,000 miners spent the winter in British Columbia.

This Fraser fever was a remarkable feature in the history of the State, and I should dislike much to leave any false impression upon the minds of my readers in regard to it. Many persons believe and have said that all the persons who went to British Columbia, during the excitement, exhibited great folly; but I never thought so. The bars on Fraser river were extremely rich, and they justified the belief that a large extent of country about the river was rich in gold. We had well authenticated accounts of gold having been found in bars for a distance of one hundred and fifty miles along the river, and it having been found in exceeding richness in some of the bars— a richness which might well compare with that of the bars of the Yuba and Feather rivers in '49. This fact of the richness of these bars, established or taken for granted, there was a reasonable presumption in the minds of all acquainted with the mines of California, that there was a rich and extensive mining country above. The gold on Fraser river is very fine; it must have been washed a considerable distance. There are therefore dry diggings up the river. The river is a very large one, and since the bars are rich, the dry diggings above must be rich and extensive. This is the universal rule in California.

It has been said that the Klamath is an exception, but the Klamath has rich and extensive dry diggings on the banks of its tributaries, the Scott and Shasta rivers. In California we do not find paying bars on our large rivers; we find none on the Sacramento or San Joaquin within three hundred miles of the ocean; and the smallest streams are, as a rule, the richest, in proportion to the amount of dust in their beds. Now Fraser river has a body of water five times—perhaps ten times—as great as that of the Sacramento, and yet Fraser river had rich bars within eighty miles of the ocean; and these facts led experienced miners to the inference that it must run through a large and rich auriferous district.

The great error, and the cause of most of the evil which has resulted from the excitement, is in regard to the supposed time when the river would fall. It was well understood, before there was any migration of note, that not until the river should fall could there be much mining; but the men familiar with the streams in Oregon and Washington Territory, supposed that of course Fraser river would rise and fall at the same time with the Columbia, and the tributaries of Puget sound, which rise in May and June, as the sun gets warm, and fall in June and July, usually getting down to a very low stage before the first of August. There seemed to be nothing unreasonable in this presumption, at first sight; no fact inconsistent with it was known; no allowance was made for the slower melting of the snows of the extreme north, at the head of Fraser river; nobody knew, by actual observation, at what time the river would fall; and so it was taken for granted by all, that the river would be low on the first of August, and would be at a fair stage for mining in the middle of July. The result proved that the river did not fall until late in September.

Many of the Fraser river adventurers ought not to have gone, even had New Caledonia been a second California; but they were no more unwise than a large portion of the Californians who habitually enter, as nearly all do, into indiscreet speculations. Many of the adventurers had become disgusted with California, had nothing to hope for here, and had nothing to lose; and, of course, it was well enough for them to go. Others were men who had $200 to $1,000, with no one to care for (to-day or to-morrow) save themselves, cut off by the unwise land system of California from the hope of having homesteads in the land upon which they have worked; and why should they not go? Take it for granted that they should lose all their money; they would see the country, and would run the chance of making a fortune, if the mines should prove as rich as those of California in '48. It was a bet of $1,000 against $20,000. Who can say that the chances were more than twenty to one against the river? And the chance of living through such another excitement as that of California in '49—the fun would be worth a fortune almost. And then after the migration had taken place, after there was a strong presumption that a similar migration would take place from

New York, England, Canada and Australia, as from San Francisco, there was then reason for believing that even if the mines were not richer than those of Colville, they must still be worked by a large population. Considerations like these justified a third class, such as storekeepers, boarding house keepers and mechanics, in starting.

Reply may be made, that they ought to have waited till the gold should have come. I do not know this. To expect such conduct, would be to expect Californians to exercise more prudence in regard to Fraser river than they do in regard to other business transactions. They are a fast people; they will attempt to outrun old Time himself, and if they succeed once, it pays them for a dozen failures. They may lament their misfortunes, and be loud in their wail about Fraser river, but if a new one were to turn up to-day, they would not wait to let anybody else get ahead of them. The cautious bump, which teaches men to wait for the gold to come before they go to new diggings, is small in the crania of Californians.

The Washoe Fever. § 31. The next great excitement among the miners of California was caused by the discovery of the silver mines at Washoe, as the argentiferous region in the basin of Carson river is called, although that name was previously given to a small adjacent valley without mineral wealth. The gold placers of Carson valley were opened in 1852, and since that year there had been a few miners permanently settled there. After exhausting the placers they found rich auriferous quartz leads, the gold of which contained a large proportion of silver. It was not until the middle of 1859, that the argentiferous veins were found, and the reports commanded little faith for some months. In October, however, the news began to attract general attention, and before winter closed up the roads across the Sierra Nevada, forty tons of ore, worth $5,000 per ton, had been sent to San Francisco, where it was smelted at a cost of four hundred and twelve dollars per ton. For four or five months, then, an excitement raged throughout California on the subject of Washoe. Many went thither, and others speculated in claims. All kinds of claims were in demand, and veins without a particle of silver or gold in them, were sold at prices varying from ten to one hundred dollars per foot. Fortunes were thus made and lost. The excitement affected the course of trade, and caused no little stringency in the money market for a time. Towards midsummer, the fever abated; worthless claims fell to their proper position, and the value of rich lodes was better understood. In August, the silver district of Esmeralda was discovered in the basin of Walker river, a hundred miles southward of Washoe, and a month later, the argentiferous veins of Coso, two hundred miles southward from Esmeralda, were found. Esmeralda and Coso have not created much excitement as yet, although several thousand persons have visited the former district.

Mining Inventions of California. § 32. The tom, the sluice, and the hydraulic process, are generally considered as Californian inventions, but they were previously known in other countries, though not so well made, or so effectively or generally used. The tom and the ground sluice had been tried in the placers of Georgia previous to 1848, but they were much improved when applied in the California placers.

Ansted, in his *Gold Seekers Manual*, (pp. 85, 86, 87) published in London in 1849, says:

"At the commencement of the mining system in the Brazils, the common method of proceeding was to open a square pit, till the workmen came to the *cascalho;* [pay dirt] this they broke up with pick-axes, and placing it in a wooden vessel, broad at the top and narrow at the bottom, exposed it to the action of running water, shaking it from side to side, till the earth was washed away and the metallic particles had subsided. Lumps of gold were often found from two and a half to twelve ounces in weight, a few of which weighed twenty-five to thirty-eight ounces, and one, it is asserted, weighed thirteen pounds; but these were isolated pieces, and the ground where they were discovered was not rich. In 1724 the method of mining had undergone considerable alteration, introduced by some natives of the northern country. Instead of opening the ground by hand, and carrying the *cascalho* thence to the water, the miners conducted water to the mining ground [ditches] and washing away the mould, [ground sluicing] broke up the *cascalho* in pits under a fall of water, [the principle of the hydraulic process] or exposed it to the same action in wooden troughs, [sluices] and thus a great expense of human labor was saved. At the commencement of the present century, there was a general complaint in Minas Geraes, that the ground was exhausted of its gold; yet it was the opinion of many scientific men, that hitherto only the surface of the earth had been scratched, and that the veins were for the most part untouched. The mining was either in the beds of the streams, or in the mountains; in process of time the rivers had changed their beds, and the miners discovered that the original beds [ancient river beds] were above the present level of the water and the banks of the streams, which formed as it were a second step, while the actual beds are the third and lowest. All these are mining grounds; the first is easily worked, because little or no waters remain there; the surface had only to be removed and then the *cascalho* was found. In the second, wheels were often required to draw off the water, while the present bed of the stream could only be worked by making a new cut, and diverting the stream."

But an admission that the general principles of all the important mining inventions of California were previously known in other countries, does not deprive the miners of the Sacramento basin of their right to high credit; for although the general principles might have been elsewhere applied at an earlier time, it was here that the inventions were brought to their highest state of efficiency, that they were universally adopted, and that they were adapted to the peculiar circumstances of every locality. It was only among a people so enterprising, so bold, so familiar with mechanics, and so dexterous in all kinds of labor, that the higher mining inventions could find immediate favor, or come into general usage. Men might be found in other nations to make the inventions, but miners would scarcely be found in any other race than the Anglo-Saxon, to at once adopt them universally. A number of machines for washing dirt are now used in the placers of Siberia, but they are worthless when compared with our Californian mining machines, and I have not thought it worth while to translate Zerrenner's description of them. (*Zerrenner's Anleitung*, p. LI.)

Exhaustion of Mines. § 33. In 1849, it was predicted that the mines would be exhausted within five years, and that the population of California would go as rapidly as they had come. This prophecy has proved untrue, but it was justified by the facts then known to the public. The success of the large body of miners now at work in California is due to processes then unknown, and to the discovery of deposits of gold, such as nobody thought of then. If our miners had now to depend altogether upon the pan, the cradle and the quicksilver machine, and knew nothing of quartz veins, the gold product would not be one-tenth of what it now is, and the population of the country would be correspondingly small.

Various changes since '49. § 34. The tom and the quicksilver machine are very rarely seen now in the mines; in some counties, neither has been used for several years. The rocker is given up almost wholly to the Chinamen. In 1849, nine-tenths of the miners were employed in working in river claims; now, such claims do not furnish employment to one-tenth of the miners. Then, the miners worked separately, or in companies of two or three; now they work in companies of five or ten. Then, it was rare to see a miner working as a hired laborer; now, at least half of the miners are hired by others. Then, men were running about continually; now, they are more permanent, because their claims are deeper, and much time must be spent in opening them. There are now in the State about six thousand miles of mining ditches, made at a cost of $15,000,000; three hundred quartz mills with three thousand stamps, costing $5,000,000, and five hundred and nineteen arastras, about one-half of which are used for amalgamating in connection with stamps, and the others for both pulverizing and amalgamating.

New Almaden. § 35. The New Almaden quicksilver mine has had a singular history. It had been known to the Indians the last century, and they went to it for the purpose of getting the decomposed cinnabar or crude vermilion as a paint. The Indians told of it to the Spanish Californians, who did not know what the mineral was. In November, 1845, a captain in the Mexican army, named Andres Castillero, visited the place, discovered the nature of the ore, laid claim to the mine, formed a company, and commenced working it. He went to Mexico soon after, and while there, sold out, and Barron, Forbes & Co., British merchants in Tepic, became the leading shareholders; and James Alexander Forbes, British Vice Consul at Yerba Buena, (now San Francisco) was placed in charge of the mine. In 1851 it began to yield largely of quicksilver. The next year, in accordance with the requirements of an act of Congress, the owners of the mine filed their claim to it before the United States Land Commission, which tribunal confirmed their title. An appeal from this decision was taken on behalf of the United States, so the case was to be tried over again. In 1857, James Alexander Forbes, who had charge of the mine from '47 to '50, came forward to prove that the title was forged, and he produced a number of letters to prove the forgery. The United States Circuit Court, in October, 1858, issued an injunction to stop the working of the mine. The owners indignantly denied the charges of forgery, and solicited the Government to allow them to take testimony in Mexico. Their application, renewed in many different forms, was denied in all. Possessing great wealth, they chartered a steamer, sent one of their attorneys with her to San Blas; he went into the city of Mexico, examined the archives of the government there, had a great number of documents copied, and returned with an ex-prime Minister, two professors of the Mining College, three persons who had been clerks in the ministries when the title was issued in 1846, and several other witnesses. The examination of these witnesses occupied months. Finally, when the case came to be submitted, the testimony which had all been printed, occupied three thousand octavo pages; the lawyers, among whom were two United States Senators, who had come from Washington for the express purpose of attending to this suit, occupied twenty days in argument, and the opinions of the judges, when printed, covered two hundred and thirty octavo pages. The Court confirmed the claim, recognized the title papers as genuine, and fastened upon witnesses for the Government the crimes of perjury and forgery which had been charged upon the claimants.

The Gold Yield. § 36. The amount of gold produced by the mines of California has never been ascertained precisely, nor is it ascertainable. Large amounts, of which no record was kept, are carried away by individuals, and the amounts were much larger previous to 1855 than they are at the present time. My estimate of the gold yield of the State is as follows:

Year	Amount	Year	Amount
1848	$10,000,000	1855	$55,000,000
1849	40,000,000	1856	55,000,000
1850	50,000,000	1857	55,000,000
1851	55,000,000	1858	50,000,000
1852	60,000,000	1859	50,000,000
1853	65,000,000	1860	45,000,000
1854	60,000,000	Total	$650,000,000

In regard to the future gold yield of the State we can only

guess. I expect that it will decrease gradually and slowly to about $30,000,000, at which figure it will stand still for many years; and the silver yield of the coast will gradually increase; but our knowledge about our silver lodes is so indefinite, that I must not even guess at the amount of their produce.

CHAPTER II.

MINERALOGY OF GOLD.

Metals obtained on the Coast. § 37. Gold, silver and quicksilver are the only metals of which mines are regularly wrought in the Pacific States of North America. Platinum, iridium and osmium are found with the gold of the placers, but they are rarely bought or sold in California. A tin mine, reported to be valuable, has been found in San Bernardino county, but no metal has yet been obtained from it. There is a rich mine of antimony twenty miles west of the Tejon Pass, but it cannot be wrought at Californian rates of labor. Rich copper mines are found in many parts of the State. No rich mines of lead, iron or zinc are known in California. There have been reports of the finding of small diamonds. The itacolumite or quartzose sandstone, however, which usually accompanies diamonds, is not found on the coast, so far as I have heard. Some other gems have been found in the placers, but none of value sufficient to pay for mining for them. It is reported that opals and rubies have been found in Table mountain, Tuolumne county; but if so, they have been few and of little value.

Coal is found at Coose Bay in Oregon, at Bellingham Bay, Washington Territory, and at various places on the shores of Vancouver Island and British Columbia. There have been numerous reports of the discovery of coal veins in various counties in California, and many thousands of dollars have been spent in opening them; yet they furnish little coal for the market, nor is it likely they ever will furnish much. The geological formation of the State renders it probable that California will always import coal. We have little or nothing of the secondary formation which bears the valuable coal deposits. The coal of Oregon, Washington and the British Colonies is tertiary, and is far inferior to the secondary coal of England and the Eastern States. California now imports nearly all her coal from New York and England.

Sulphur lies in deep and extensive beds about Clear Lake, in Napa and Sonoma counties, and in Santa Barbara county.

Borax is found at Borax Lake, in Napa county.

Asphaltum is obtained in many counties along the southern coast of California, but it is dug up from the surface, without any labor that deserves the name of mining. The asphaltum is formed by the drying or hardening of a mineral oil that rises from numerous springs.

No Ore of Gold. § 38. Gold is the only yellow metal. It is always found in nature in a metallic state, never as an ore. Webster says an ore is "the compound of a metal and some other substance, as oxygen, sulphur or carbon, called its 'mineralizer,' by which its properties are disguised or lost." I would define an ore to be "a natural compound, containing a metal (from which it differs in appearance) in chemical union with one or more non-metallic substances." Gold is found in a state of nature combined with other substances mechanically, not chemically. Gold mines are of two kinds, placer and quartz.

Placer Mines. § 39. In placer mines, gold is found in earthy matter, where it has been deposited among clay, sand and gravel under the influence of water. The word "placer" is Spanish, and in its first signification means pleasure. I apply the name to all diggings where gold is found in diluvium or alluvium, without regard to their depth. In 1849, the term as used by Californians signified diggings on the surface of the ground, or within a few feet of the surface, because deep diggings were then unknown. All placers, even if but a few inches deep on the surface of the ground, are called "mines;" though that word, as used by those engaged in searching for other minerals, implies deep digging into the earth.

Quartz Mines. § 40. In quartz mines, gold is found in stony veins. These veins are usually quartz, though sometimes the metal is found in granite, limestone, talcose slate, greenstone, serpentine, porphyry, trachyte and trap. It is supposed, however, that these rocks are auriferous only where they have come, while fused, into contact with auriferous quartz. Much auriferous metamorphic limestone and granite has been found in California, but nobody ever speaks of "limestone mining," or "granite mining;" all persons who get their living by mining for gold, are either "quartz" or "placer" miners.

Forms of Quartz Gold. § 41. Gold has many different forms in quartz. In some lodes, the particles are all so small as to be barely perceptible to the naked eye; in others, they are flat and small; again, they may be lumpish, varying in size from a pin-head to a pea; elsewhere, in smooth sheets, as thin as paper and smooth as glass, though enclosed between rough rock on both sides; here, like fern leaves; there, in threads; in this vein, dendritic, resembling a tree or coral, with a multitude of branches shooting out from a common center, and every part made of a little golden octahedral crystal; in that vein every particle of gold is independent of its neighbor; in another, all the particles are connected together, as though when the quartz had been in a state of fusion, the gold had been poured in and stirred about a little, (but not enough to break the connection) until the rock had solidified. Some very beautiful specimens of quartz gold, resembling fern leaves, have been found near Shingle Springs, in Eldorado county. Elegant samples of dendritic gold have been obtained near Coulterville, Mariposa county. "The most beautiful specimens of crystalline gold," says Blake, (*U. S. Pacific Railroad Report*, vol. v, p. 300) "are those in which the crystals are combined with an arborescent or dendritic growth of the metal, like the leaves of ferns or foliage of the arbor-vitæ. Specimens of this character are not found among the rudely transported drift, but can be obtained only from their original bed in the solid rocks. Some of the most remarkable and beautiful specimens ever seen were taken out of the cavities of a quartz vein at Irish Creek, about three miles from Coloma. They presented various and complicated combinations, being arborescent, in broad, paper-like sheets, studded with brilliant crystals, and in solid octahedrons. These were combined together in the most interesting manner, giving an effect beyond the reach of art. A very fine specimen of this character, in my collection, has the form of a leaf; one side is aborescent and very brilliant, and the other is studded with about twenty-five perfect octahedral crystals. They are geometrically arranged, all the similar edges being parallel. This is believed to be the most beautiful and curious specimen known. Its weight is seventeen pennyweights and ten grains; length, two and a quarter inches: width, one and a half. One of the largest specimens of this arborescent and foliated gold, from Irish Creek, was about twelve inches long by twelve broad. A part of the specimen was a plate three or four inches long, covered with triangular marks; the remainder was arborescent, and the whole appeared to have grown from one end. Another specimen, slightly different in character, and probably from another locality in the vicinity, was ten inches long, three broad, and about half an inch thick. It weighed thirty-one ounces, and was free from quartz—forming a most beautiful mass, of a rich yellow color, and a delicately marked surface, consisting of a network of fibers. It appeared like a bundle of fern leaves, closely matted together. With one exception, all the crystals which have come under my observation are octahedrons; not a single cube has been seen. The exception is a large crystal, a pentagonal dodecahedron, upon one of the broad plates of gold." Very few auriferous quartz veins in the State furnish rock hard enough for quartz jewelry. In most lodes, the rock, especially in the richest places, has little fissures and cavities, and is ready to crumble to pieces at any slight blow. The Marble Spring lode, in Mariposa, has furnished more rich quartz, suitable for jewelry, than any other vein in the State. The solid lumps of gold found in quartz, never, as far as I have heard, weigh more than two ounces, and, indeed, a compact mass of gold, weighing an ounce, is more rare in quartz, than lumps weighing ten pounds are in the placers.

Gold Dust. § 42. Placer gold is usually called "gold dust;" but the word dust, without explanation, conveys an erroneous idea. Gold dust is not a fine powder, but ordinarily consists of pieces larger than a pin-head, very often with lumps varying from a pennyweight to an ounce; and, even if the lumps weigh ounces or pounds, it is none the less "gold dust."

Forms of Placer Gold. § 43. The first general division of placer gold, or gold dust, relates to the size of the particles, and is denominated "fine" and "coarse." Fine gold dust is that whereof the particles are smaller than pigeon-shot; in coarse gold, the particles are larger. Fine gold is often found without any admixture of coarse; but the latter is rarely found without containing many small particles. Coarse gold is usually made up of pieces very irregular in size and shape.

"Fine gold" may be divided into "flour," "grain," "shot," "fine shot," "scale," "thread," and "spangle."

"Flour gold," is as fine as wheaten flour.

"Grain gold," is that whereof the particles are not larger than the grains of gunpowder.

"Fine shot," is that whereof the particles are of a roundish shape, and not larger than the smallest shot. Fine shot gold was found in Secret Ravine, Placer county.

"Shot" differs from "fine shot" only in size, being a little larger.

The particles of "scale gold" are flat, like little scales. They usually have a shape approaching the circular, and are not more than an eighth of an inch in diameter. In many places the scales are remarkably uniform in size and shape. Thus, the gold dug on the bars of the Yuba and Feather rivers, in '48, '49 and '50, was most of it in scales, nearly circular, about a tenth of an inch in diameter, and a thirteenth of an inch in thickness. Similar scale gold has been found in ravines in nearly every auriferous county of California.

"Thread gold" has the shape of threads, about as thick as a large pin, and from a quarter of an inch to an inch in length. Sometimes the longer pieces are found knotted up very curiously and inexplicably. Thread gold is found in Fine Gold Gulch, Mariposa county, and also in several gullies near Camptonville and Yreka. In some places the threads have a groove running through them, so that they are semi-cylinders; in other places the threads are fluted; and the same general character is usually observed in all the threads obtained from one gully. In silver mines it is not unusual to find little bunches of threads of native silver, all tangled together. These bunches suggest the idea that probably the gold threads were all in similar bunches. Thread gold is usually of a very low fineness—that is, it contains a large proportion of silver.

"Spangle gold" borders on the coarse; in fact, there are "fine spangle" and "coarse spangle." "Spangle gold" has shapes similar to those which are taken by lead when it is poured in a melted condition from a hight of four or five feet, and allowed to fall into water or upon damp ground. The particles are of many fantastic shapes. There is a "Spangle Gold Gulch," near the head of the San Joaquin river.

Coarse gold may be divided into "crystalline," "coarse shot," "pea," "bean," "cucumber seed," "pumpkin seed," "moccasin," and "miscellaneous coarse."

"Crystalline gold," in octahedral crystals, is very rarely found in placers and not often in quartz.

"Coarse shot gold," is that whereof many or most of the particles are roundish in shape, and larger than pigeon-shot in size.

"Pea gold" is made up of roundish lumps, about the size of a pea. Pea gold was found in several ravines, emptying into Cottonwood Creek, in Shasta county.

"Bean gold" has smooth lumps, resembling a bean in size and shape. Samples of it were found near Horsetown.

"Cucumber seed gold" resembles cucumber seeds, and sometimes the resemblance of the lumps to the seeds, and the uniformity of size and shape, is very wonderful. Cucumber seed gold was found in various cañons of Clear Creek, and on some bars of the Mokelumne river.

"Pumpkin seed gold" resembles pumpkin seeds.

"Moccasin gold" is in lumps, shaped like a moccasin or low shoe, and about half an inch in length. Moccasin gold is found in Coarse Gold gulch, in Fresno county.

The "miscellaneous coarse gold" makes up almost the entire amount of coarse gold taken from the mines. It is composed of particles of irregular size, usually flattish in shape, and smooth. The other kinds of coarse gold and the fine thread were not abundant in 1849, at the time when the placers had just been fairly opened; but now they are very rarely found.

I know of no satisfactory explanation for the fact that the particles of gold in auriferous quartz never approach in size the nuggets found in adjacent placers. There are only two theories. One, that the auriferous quartz was much richer and had the gold in larger masses in those parts of the veins that have been worn down, than in those that still preserve their original form. The other theory is, that the numerous small particles of gold lying near each other in the placers, are collected together by some chemical or electrical influence and united into one mass. (*Gangstudien*, vol. 3, p. 464.)

Large Nuggets. § 44. There is some doubt about the size of the largest lump or nugget of native gold. According to a legend current in Peru, a nugget weighing 400 pounds was found at the mine of San Juan de Oro, on the headwaters of the Amazon river, in the time of Charles the Fifth, and was sent as a gift to that monarch, who rewarded the donors by elevating them all to nobility. I have not been able, however, to find any trustworthy authority for this legend.

The largest nugget of which we have an unquestioned account is the "Welcome nugget," weighing 184 pounds Troy, and worth $42,000, found at Ballaarat, Australia, on the ninth of June, 1858. It was found 190 feet below the surface of the earth, in a claim from which many other nuggets, weighing from ten to forty-five ounces had previously been taken. The Melbourne *Herald* of the eighteenth of June, described it thus:

"A large, misshapen, irregular lump of gold, water-worn and rounded upon each of the numerous edges presented by a surface completely and more or less deeply honeycombed. Its total length is about twenty inches, its greatest breadth about twelve inches, and its greatest depth about eight inches."

The second nugget in size was found in Calaveras county, California, in November, 1854. It weighed 161 pounds avoirdupois, and was supposed to contain twenty pounds of quartz, and was valued at $29,000. The value was less in proportion to the weight than that of the Welcome nugget, because the latter was free from quartz and dirt, and because the Australian gold has less silver than the Californian.

The third nugget in size was found at Korong, Australia, in the summer of 1857. It was called the "Blanche Barkly nugget," weighed 145 pounds, and was estimated to be worth about $35,000. A newspaper described it as "a solid mass of virgin gold, two feet, four inches long, ten inches broad, and from one to two inches thick."

It is reported that a nugget weighing ninety-three pounds was found in 1842, in the Valley of Taschku-Targanka, in Siberia.

A nugget of eighty pounds was found in Cabarras county, North Carolina, in 1842. (*Gangstudien*, vol. 3, p. 468.)

Many boulders of rich auriferous quartz, varying from 100 to 500 pounds, have been found, but these do not deserve to be classed among the nuggets.

Probably not less than a thousand nuggets, weighing a pound or more each, have been found in California. There have been not less than fifty weighing over twenty pounds each. The number of those weighing from one to six ounces, would be almost innumerable. The richest place in California for nuggets, was the vicinity of Sonora, in Tuolumne county. The following is an incomplete list of the nuggets found there from 1850 to 1858, inclusive. The dates up to March, 1854, refer to the daily *Alta California* newspaper, in which the nugget is mentioned; the subsequent dates mentioned, merely the month in which the lump was found. In some cases the weight, in others the estimated value is given:

1850.

23 ℔s. Wood's Diggings	February 20	10 ℔s. 11 ozs. Sonora	May 14
5 ℔s. near Sonora	March 6	18 ℔s. Sonora	June 7
51 ozs. Sonora	April 2	4 ℔s. 4½ ozs. Jamestown	August 11
23 ℔s. 2 ozs. Sonora	May 14	13 ℔s. Sonora	October 14

1851.

28 ℔s. 4 ozs. Sonora	October 5	23 ℔s. 6 ozs. near Sonora	October 5
24 ℔s. Sonora	October 5	69½ ozs. Wood's Creek	December 1

1852.

$90, Sonora	January 5	$80, Sonora	January 10
12 ozs. Sonora	January 5	26 ozs. Shaw's Flat	August 55
$1,100, Sonora	January 10	116 ozs. Shaw's Flat	November 23
$900, Sonora	January 10		

1853.

29 ozs. near Sonora	January 18	15 ozs. Columbia	May 13
20 ℔s. 7 ozs. near Sonora	February 19	11 ozs. Columbia	May 13
$1,500, near Sonora	February 21	9 ℔s. Indian Gulch	May 16
9 ozs. near Sonora	February 24	7 ℔s. 8 ozs. Indian Gulch	May 16
7 ℔s. near Sonora	February 25	36 ozs. Yankee Hill	June 5
69 ozs. near Columbia	February 26	12 ozs. Shaw's Flat	June 12
7 ozs. near Sonora	March 4	4½ ozs. Shaw's Flat	June 13
116 ozs. Columbia	May 2	30 ozs. Sonora	June 29
24 ozs Columbia	May 2	71 ozs. Sonora	June 29
18 ozs. Columbia	May 13		

1854.

11½ ozs. Sonora	February 11	2 ℔s. near Columbia	June —
27 ℔s. Columbia	March 23	16¼ ℔s. Sonora	July —
1 ℔. Jamestown	June —	7½ ℔s. near Columbia	September —
$400, Springfield	June —	17 ℔s. Sonora	November —

1855.

30 ℔s. near Sonora			January —

1858.

41 ozs. Columbia	May —	15 ozs. Columbia	September —
13 ozs. Columbia	May —	33½ ℔s. Columbia	September —
11 ozs. Saw Mill Flat	May —	33 ozs. Columbia	September —
47 ozs. Columbia	July —		

CHAPTER III.

CHEMISTRY OF GOLD.

Chemical Fineness of Gold. § 45. Having considered gold mineralogically, that is, in regard to the shapes in which it is found, we now come to consider it chemically. When pure, it is of a bright yellow color, nineteen times heavier than water of the same bulk, heavier than any other metal in common use, and fusible at about 2800 deg. of Fahrenheit's thermometer. Gold, as found in nature, is never pure; it is always mixed with silver, frequently with copper, and sometimes with other metals. We do not know why it is always found in combination with silver. The proportion of silver varies greatly, from almost pure gold with a little silver in it, to almost pure silver with a little gold in it. The silver and other metals in native gold are called "alloy." The proportion of pure gold in native gold is designated by the word "fineness;" the greater the proportion of pure gold, the higher the fineness. The word "fine," as applied to gold dust, has, therefore, two distinct significations—one mechanical, referring to the size of the particles; the other chemical, referring to their chemical composition. Jewellers usually estimate the fineness of gold in carats, of which twenty-four make an ounce; and perfectly pure gold is said to be twenty-four carats fine; whereas, if there be two carats of alloy in an ounce of the metal, it is twenty-two carats fine; or if there be three carats of alloy, then it is twenty-one carats fine, and so on. But at the Mint, and in the common usage of California, the fineness of gold is described in thousandths. If fifty parts in a thousand are alloy, then the metal is said to be nine hundred and fifty fine. Our standard coin contains ninety parts of copper, ten of silver in a thousand, making one hundred parts of alloy altogether, and, therefore, it is nine hundred fine. Pure gold is a thousand fine. The gold of the Californian placers varies from five hundred to nine hundred and ninety-five fine, but the latter fineness is extremely rare. A reasonable approximation to the fineness of gold can ordinarily be made by its color, which varies according to the amount of silver, the usual alloy. Our gold coin is redder than pure gold, because it is nearly one-tenth copper; the placer gold of California has a lighter color than pure gold, because it has about one-tenth or more of silver.

Mr. Molitor, in an essay on gold published in the *Alta California*, said:

"About 1853, a gold specimen, of the size of a man's hand, found somewhere in the neighborhood of Downieville, (according to the statement of the depositor) was assayed in the laboratory of the late firm of Wass, Molitor & Co., and found to be nine hundred and ninty-two fine. This was an unique case; but gold of above nine hundred and seventy fineness has been frequently assayed in San Francisco. On the other hand, the gold from the Kern river mines contains such a large proportion of silver, as to be almost identical with the Electrum of the ancients, or the Zoroche of the Mexicans, which means, a metal consisting of about half and half, silver and gold. Between these two extremes all degrees of mixture of the two metals have been found in this country. The experience of several years shows, however, that eight hundred and fifty-five would be about the average fineness of California gold, to which it must be added, that by far the greater part of the whole gold produce seems to group itself, in regard to fineness, close around this average figure. On the virtue of this statement we may say, therefore, that the greatest part of the gold of this country ranges, as a rule, between eight hundred and forty and nine hundred and thirty, and that all cases exceeding these limits may be regarded as exceptions of the general rule.

"It is impossible, even to the most practised eye, to determine the quality of an unknown sort of gold dust by merely looking at it; and even in judging a well known description of dust, the purchaser may deceive himself very easily, to his own damage. The gold may, for instance, by some natural accident, possess a richer color than entitled to by its quality; or it may be taken for a superior kind of gold, on account of the shape of its grains, which may be similar to some known dust of good quality; or it may be mixed with some inferior gold, either with or without an intention to defraud the buyer; or adulterated in some way or another, and so on.

"Even the knowledge of the region, or gold field, from where a certain description of gold originated, is not a sure evidence of its quality. Nobody can depend on it, that the gold taken out of one and the same flat, hill, bar, or even from the same claim, or quartz lead, will always be exactly the same. Very often the most astonishing differences in this regard are found within comparatively short distances. Thus, there are quartz leads with very low gold, surrounded by placers famous for the fineness of their metal; and on the contrary, veins with very rich metal in the vicinity of diggings not renowned for the superior quality of their metal.

"There is, in fact, one sure method to determine the fineness, and consequently the exact value of the precious metal, and that is the regular metallurgic process of assaying, after the previous melting of the dust into a bar or ingot.

"1. Gold coming from British Columbia, or the Fraser river mines, generally ranges between eight hundred and forty and eight hundred and sixty fine. In some cases it was found as low as eight hundred and twenty; in others, above eight hundred and sixty; but these may be considered as exceptions to the rule. It mostly appears in our market as coarse lumps of amalgam gold, and suffers an average loss of ten per cent. by melting.

"2. The average fineness of dust from the Gold Beach, above and below Port Orford, (Oregon) is eight hundred and eighty. The gold dust appears throughout in fine scales, and is extracted from the sand and accompanying minerals, including Iridio-Platinum, chiefly by amalgamation.

"3. The gold which finds its way to San Francisco principally by Crescent City, and, therefore, has been worked chiefly on the Klamath river and its tributaries, seldom exceeds eight hundred and eighty fine, and seldom descends below eight hundred and fifty. Its average fineness is eight hundred and sixty-five. In this district, we include the counties of Del Norte, Klamath and Siskiyou, and the adjoining southern border-tract of Oregon. This gold mostly appears in coarse and heavy grains, and sometimes contains a considerable admixture of Iridium.

"4. The placers on Trinity river, and on the western tributaries of the upper Sacramento, belonging to Trinity and Shasta counties, seem in general to yield a better quantity, and we may safely put the average ten thousandths higher than under the previous number. Some dust from the neighborhood of Weaverville shows the fineness of above nine hundred.

"5. Feather river gold shows an average fineness of eight hundred and ninety, and most frequently occurs in very regularly shaped and almost uniform grains or scales.

"6. Gold on the north forks of the Yuba is generally much finer than the above, in many cases going up as high as nine hundred and fifty, and seldom below nine hundred; average about nine hundred and twenty. The dust is also mostly of a scaly description, and a great deal of it appears in the market as amalgam gold.

"7. On the south fork of the Yuba, the general fineness seems again to decrease. Around Nevada, placer gold seldom shows more than eight hundred and eighty. The quartz gold from the various veins of Grass Valley ranges between eight hundred and eight hundred and fifty, and may be put down at eight hundred and twenty, average fineness.

"8. On the north and middle forks of the American river, gold is again rising in fineness, especially in the diggings about Auburn, approaching here the figure of nine hundred.

"9. On the south fork of the same river, in the vicinity of the towns of Coloma and Placerville, the fineness of the dust varies very much. Coloma gold seldom ranges above eight hundred and ninety, and generally comes nearer to eight hundred and seventy. But in the neighborhood of Placerville the gold rises, in most cases, up to nine hundred, and in some places thereabout much higher still. At Coon Hollow, a peculiar kind of dust of a dark, rusty appearance is found, which is over nine hundred and forty fine.

"10. In Amador county, around Drytown, Jackson and Volcano, the fineness of gold is rather below the general average.

"11. In Calaveras county, great varieties occur in this respect. Mokelumne Hill gold is seldom above eight hundred and ninety; San Andres averages eight hundred and ninety; Campo Seco, nine hundred and five; Vallecito rises up to nine hundred and ten to nine hundred and twenty.

"12. Tuolumne is the county most renowned for the fineness of its gold. Sonora and Columbia dust seldom falls below nine hundred, and often rises above nine hundred and fifty.

The average may be marked down at nine hundred and thirty. This gold is generally rough and coarse grained, and of a very rich color.

"13. In the adjoining county, Mariposa, the fineness of the precious metal decreases very sensibly; the average can scarcely be put higher than eight hundred and fifty.

"14. Still farther south, on the upper San Joaquin and its first tributaries, the rivers Chowchilla and Fresno, the fineness of the gold falls below eight hundred, and sometimes even as low as seven hundred. This dust consists generally of diminutive spangles, of a treacherously rich appearance, intermixed with curiously elongated, almost needle-shaped grains.

"15. The lowest degree of fineness of gold in this State is found in the most southern parts, on the diggings of Kern river and its numerous branches. This gold seldom reaches above seven hundred, and often falls down to near six hundred. Its average fineness may be fixed at six hundred and sixty.

"16. Carson Valley dust, on the eastern slope of the Sierra Nevada, although beautiful to the eye, is also exceedingly low —generally below eight hundred.

"17. Gila and Columbia river gold is of a very small description, with grains similar to Australian gold. Some parcels of it have shown the fineness of above nine hundred and seventy; others fell below nine hundred and twenty."

CHAPTER IV.

THE GEOLOGY OF GOLD.

The Formation of Gold. § 46. It has already been stated, that all the placer gold came from quartz veins which have been broken up or disintegrated. We do not know how the gold got into the quartz, but we have surmises. Our ignorance is not so gross as that of the miners of California, as a body, in 1849, who supposed the precious metal had been thrown out by a volcano or some mysterious source, and that some lucky individual would find the "fountain head" of the gold, and get enough of the metal to load a dozen ships. This erroneous impression led many an enterprising miner to give credit to wild stories, and to desert diggings that paid from twenty to fifty dollars per day, in the hope of finding the "fountain head." Although our present knowledge lacks completeness, yet we know enough now to perceive the absurdity of those ideas of 1849.

Quartz the Mother of Gold. § 47. According to common language, "quartz is the mother of gold;" in other words, all native gold is or was encased in quartz. The gold of the placers has been set free by the breaking up of quartz; the gold found in granite, serpentine, limestone, or other rock, is supposed to have been communicated from adjacent quartz, when both were in a state of fusion. We do not know why gold is found originally only in quartz, nor how it got there. There is no theory known to me about the *why;* there are several about the *how.*

The Igneous Theory. § 48. The first theory to explain the manner in which the gold in auriferous quartz veins got into the rock is, that quartz was forced up in a liquid state from the molten center of the earth, and then contained the gold now found there. In favor of this theory it is argued, that quartz is igneous in its whole character; that it is never found lying flat like aqueous rocks generally; that it has none of the stratification, peculiar crystallization, or petrifactions, found in rocks deposited in water; that it is formed in deep, narrow veins, like eruptive rocks generally; that if it be igneous, that fact would almost, of itself, prove the truth of this theory; that auriferous quartz is sometimes so compact and hard that the gold could not have been introduced into it in any other manner than when both were in a state of fusion; that sometimes two veins of quartz have been observed to cross each other, one auriferous, and the other barren in its general character; the latter is usually auriferous, too, for some distance on one side of the intersection, as though the barren stream had flowed through the rich one when both were in a state of fusion; that where auriferous quartz veins dip, as they usually do, the lode is richest on the lower side, next the foot wall, as though the heavier metal had settled down; that in those places in the lode where this streak is richest, quartz over it is the poorest; and, finally, that no other theory has so much evidence in its favor. On the other side it is argued, that the heat necessary to melt quartz would drive gold off in a vapor; that if it were not volatilized, being eight times heavier than quartz, it would necessarily sink to the bottom of the fused quartz; whereas, it is never found collected in large masses, but often in small particles, very evenly distributed through the rock; that the quartz being the first to harden, would shoot its crystals through the gold, which in fact it never has done, whereas the gold not unfrequently shoots its crystals through the interstices of the quartz; that compact quartz is almost invariably poor in gold; and that rich quartz usually contains, about the gold, much sulphurets of iron and copper, which appear to have been deposited there at the same time with the gold, and yet could not have been there under any high heat, which would decompose them and drive them off.

The Vapor Theory. ₰ 49. The second theory is, that the gold was precipitated from a vapor under electrical or chemical influences. As evidences for this theory are, that rich auriferous rock is usually full of crevices and fissures, so that vapors could pass through it; that the sulphurets accompanying gold must have been deposited from vapors, which might also have deposited the gold; and that auriferous quartz is always richest near the surface, as though the metal had not been deposited by the vapor until it approached the air and coolness. Another evidence of this theory is found in an assertion that a hole drilled into auriferous quartz in the Ural, and intended to be used for blasting, but never so used, was examined forty years after it was made, and found to be full of fine crystals of gold. Nearly every argument against the first theory is used in favor of this, and nearly every argument for that is used against this.

The Aqueous Theory. § 50. The third theory is, that auriferous quartz was formed under aqueous influences, and that the gold was deposited or precipitated at the same time with the quartz. It is known that quartz can be dissolved in water heated to a high heat, under a heavy pressure.

These are the theories, but neither one commands the confident belief of geologists.

Overman says: (*Treatise on Metallurgy*, p. 50) "all the veins and masses which do not run parallel with the strata of rock, it may be assumed, are filled rents. With regard to the manner in which the rent has been filled, different forces have been acting, and the nature of the deposit assumes, accordingly, a different aspect. Lodes which are wide at the top, with smooth walls of the same material on both sides, we are justified in assuming to be wedge-shaped, thinning gradually in the convergence of their walls. The mineral and foreign matters having been introduced from the surface of the ground, have been carried along by a current of water. Are the walls of a vein rough, and do they show signs of having been under the influence of a higher heat than the surrounding rock generally? we are warranted to conclude that the rent has been caused and filled by an expansive force from below. In the latter case, we expect an increase of mineral with the depth; and in the first, a decrease of it. Since the bulk of mineral veins is composed of sulphurets, and these are volatile, we conclude that all fissures, pockets and cavities, which are filled by sulphurets, have been so filled by the vapors of those metals deposited in the cavities. The lead ores of Missouri and Arkansas owe their origin to this cause; also the gold ores of the Southern States, and, in fact, most of the pyriteous ores of the Eastern States."

The "Country" of Gold. § 51. Auriferous quartz is always found near granite and slate. It is useless to seek for gold in places where the fundamental granite or "bed-rock" of the earth is covered, thousands of feet deep, with aqueous rocks, such as the coal beds of Pennsylvania or the blue limestone of south-western Ohio. But where quartz, slate and granite are found together, there gold will probably be also. The Sierra Nevada mountains are chiefly of granite, with some slate, trap, trachyte, serpentine and metamorphic limestone, in the gold region; and near the level of the Sacramento basin, some sandstone.

Rules of Quartz Veins. § 52. Auriferous quartz veins, in California, have usually a direction from north-northwest to south-southeast, and a steep dip to the eastward. The thickness varies from a line to fifty feet. The quartz is generally white, or bluish-white in color, with rusty streaks or spots, caused by the decomposition of iron pyrites. The gold is never equally diffused through the rock; some parts of the vein are always richer than others. Quartz containing two ounces of gold to a pound of rock is very rich, and is a rarity, or at least a subject of admiration among all quartz miners, and in every district. The richest parts of a vein are in streaks. There is almost invariably a rich streak along the foot-wall, or in the lower side of the lode, and this is often the only part of a vein that will pay for working. All the gold of a lode has a uniform character; that is, the particles bear a general resemblance to each other, in size, and in chemical fineness. There are exceptions, however. Sometimes a lode will have several rich streaks, one of coarse particles, another of fine, and of different chemical fineness. Lodes are found, too, in which the rich streaks are not parallel with each other, nor with the plane of the vein. Nearly all rocks have some kind of a grain, cleavage or stratification, and metalliferous veins run with or across the direction of this cleavage. It is a general rule that the most extensive veins are those which run across the cleavage of the "country," or "bed-rock." The veins which follow the plane of cleavage of the "bed-rock" are often small deposits, which, though they may be richer than the transverse lodes, are yet soon exhausted. Another general rule is, that metalliferous lodes are richest where several branches of the vein, after having spread out, unite again, and are compressed within a narrow space. A third rule is, that lodes are usually rich where they cross from one rock to another, and where they have one kind of rock for the hanging-wall and another kind for the foot-wall. Marcou (*Geology of North America*, p. 82) says the richest quartz veins of California are found where sienitic granite and trap meet.

Quartz Veins Poorer as they Descend. § 53. It is the general opinion of the quartz miners of California, that auriferous quartz leads become poorer as they leave the surface. This, also, is the opinion of miners and mineralogists who know something of gold mining in other countries. (*Zerrenner Anleitung*, p. 10; *Zerrenner Russlands Bergwerk's Production*, p. 18; *Oscar Lieber, Gangstudien*, vol. III, p. 506.) The last-named authority states that in one lead in North Carolina, the gold near the surface was of a higher chemical fineness than that deep down; and that the line dividing the rich from the poor metal could be distinctly traced with the naked eye, by the difference between the color of the metal above and below. It has been asserted, too, that the gold obtained from certain quartz claims in California is not worth so much per ounce, as gold taken from the same claims eight years ago, and nearer the surface. There are numerous rich lodes of auriferous quartz in Washoe, and the opinion is frequently expressed, that, after going down two or three hundred feet, the gold will probably disappear, and its place will be supplied by silver. Veins of auriferous quartz are often wider below the ground than at the surface, and some persons say that in such cases, a square yard cut across the vein, near the top, will contain as much gold as a square yard cut across deep down, where the vein is twice as wide, and where, consequently, the amount of rock is twice as great. I do not consider these ideas, however, to be proved, fully. Some persons who have seen much of quartz mining, express the opinion that quartz leads are as rich, deep in the ground, as near the surface.

The Great Quartz Vein of California. § 54. Mr. S. V. Blakeslee has written thus of the auriferous quartz of the Sierra Nevada:

"To appearance, no order or system prevails in the distribution of the gold-bearing veins scattered here and there through all parts of the mining regions; but, evidently, a thorough geological survey of the State might reduce them to a fair system of trunks and branches. Yet, a little observation of their localities will reveal one great leading vein, taking precedence of all others, running along the Sierra Nevada, lying about an average distance of sixty miles west of their summits; and although broken at intervals, and irregular, yet extending along a good portion of the length of the State. Vast quartz mining operations are already going on at different points, and we do not doubt that this vein will in time yield more gold than the present circulating currency of the world. It deserves a distinctive name, as much so as the great rivers, seas, and mountains of the earth; but at present we can refer to it only as the Great Gold-bearing Quartz Vein of California.

"In thickness, it varies from two to twelve or sixteen feet"; yet in a few places it deepens to many rods. On each side are commonly numberless accompanying smaller veins, diminishing in thickness to a mere line. The great vein is generally destitute of more than a trace of gold, while side veins, only a few rods from it, may be immensely rich, supported by the great vein as the fruit-bearing branches of a tree are supported by the trunk. The great vein is generally composed of solid, white, crystalized quartz; while the side veins are of every variety of feature, varying to a light porous ore of iron, or soft, red, clay-like material. This, on being carefully washed, will often yield a large amount of the brightest, finest, and most beautiful dust of gold, while in other instances, the gold appears left by the disintegrated vein as a pure, porous mass of spangles, balls, and shapeless forms, just adhering to each other where they touch. Yet this last is very uncommon, as the interstices between the particles are generally filled with quartz.

"We have personally visited this vein, only from Coulterville, Mariposa county, thence north through Tuolumne, Calaveras, Amador, El Dorado, Placer, Nevada, and Butte counties; yet to the south, at least, we know it extends indefinitely. The vein is not always evident above ground. In some places it disappears for miles under the surface, and again appears; in other places it is lost for only a few rods. It especially appears at ravines, or other depressions in the surface. It descends under the ravines, but appears capping hills and mountains, towering in sharp points, at some places five hundred and a thousand feet high. Thus, just north of Coulterville, this vein can be seen from higher elevations, whitening the scene with peaked elevations for ten miles to the north. It then disappears under the Tuolumne river cañon. In three miles it again appears at Algerine ; again disappearing under Sullivan Creek, it reappears at Poverty Hill, passing through and capping a mountain four hundred feet in hight ; then continues near seven miles by Jamestown. It next descends under the cañon of Stanislaus, and suddenly rises on the north, capping Carson Hill twelve hundred feet high; then continues by Albany Flat and Angel's Camp. From here it seems to disappear for a number of miles, occasionally exhibiting traces of itself by Mokelumne Hill, to the north side of the cañon of the Mokelumne river, where it again appears in distinct form at Butte City, Jackson, Sutter Creek, Amador and Drytown. Beyond this, it cannot as yet be so easily and distinctly traced ; but we are confident that, ultimately, the rich veins in the region of Placerville, El Dorado county, a few near Auburn, Placer county, those of Grass Valley, Nevada county, and Frenchtown, Butte county, will all be found to belong to this same great vein, appearing and disappearing at different places, but really continuous under ground."

Formation of the Placers. ₰ 55. It is supposed that at some period, extremely remote in the past, the whole surface of the earth was rock, which was worn away by the action of the air and water, and became loam, clay, sand, gravel and so forth, which were deposited under water, and then in places raised by subterranean forces into the open air, and worn into their present shapes of hills, valleys, ravines and gullies by the action of other waters. We see, for instance, that many of the hills and mountains along the eastern side of the Sacramento basin, are made up of gravel evidently worn round in water, by long rubbing against other pieces of gravel, sand and clay, all lying in layers, evidently deposited regularly, at the same time and in the same manner, over a large extent of country. There were once no valleys between these hills ; the present irregularity of surface has all been caused by streams of water, which have carried away large portions of the earth and gravel previously deposited there. In the gold region, the auriferous quartz was broken up and worn away like other rock; the light material of the quartz was carried far away by the water; the heavy gold did not move far from the original position of its original quartz vein. It did move, however; it was carried down steep hill-sides into the beds of gullies, ravines, brooks and rivers. The pieces of gold were rough and irregular in shape when they were freed from the quartz; but as they were carried along by the water and rubbed against rolling stones and gravel, the angles were rounded off; and those carried very far were smashed up and broken into little pieces, and converted into flakes, scales or grains. Coarse gold is never found far from the place of its maternal quartz; the gold on the bars of large streams is usually fine, and sometimes fine gold is found in very short ravines.

In 1849, Californian miners divided the gold dust into the two general divisions of "river" and "gulch." "River gold" is usually fine and smooth; "gulch gold" coarse and rough. The particles of gold in auriferous quartz are irregular in shape, rough, and full of sharp angles; after they are liberated from the rock, and washed down the gullies by water, they are smashed flat, broken into pieces, and worn smooth. The different forms of placer gold often appear within very short distances of each other. Thus, I have known two short gullies, parallel with each other, and not more than one hundred yards apart, to have different kinds of dust—one coarse, the other fine.

Near Prairie City, in El Dorado county, there is a hill several miles long, on one side of which all the gold is round and shot-like, and on the other side it is all in scales.

Persons experienced in buying gold dust, can often tell by the size and shape of the particles—or could six or eight years ago, when most of the gold was from the placers—whence it came. Knowing the place, they could determine the value, knowing what gold from the same place had previously assayed. The color of dust depends much on the material in which it has been imbedded, and this color often decieves inexperienced gold buyers. A jeweler can guess very near the fineness of golden jewelry by its color; but there is much more danger of error in examining gold dust. After it is melted, more reliance can be placed upon the color.

Near the Tuolumne river, there is a "black lead," from which all the gold was coarse, and covered with a black covering resembling enamel, which adhered to the metal very stubbornly. It was removed by heat, rubbing and washing in strong lye.

Diluvial and Alluvial Placers. § 56. Placer gold diggings may be divided into two classes, the "diluvial" and the "alluvial." I call those diggings diluvial where the auriferous dirt was deposited in a large body of standing water, and simultaneously over a large district, as is the case with many of the hills and flats in the State. It is evident that the great body of the auriferous dirt in some of the largest counties, was deposited at one time in a sea or large lake. I call all these deposits diluvial, as if made under the influence of a deluge. That these deposits were made in a very remote age, is evident from many facts. The bones of mastodons and other gigantic animals of species long extinct, are often found in them. The deposits are found spread with similar strata over large districts, and in positions where they could not have been placed by the present streams of the country. "Alluvial" applies to the deposits made under the influence of rivers; "diluvial" to deposits made when the whole country was under water. The alluvial placers have been formed where streams of water have run through these diluvial deposits after they were raised to the air, have swept away much of the dirt, and have changed the position of the gold. The diluvial diggings usually lie deep, and are buried under many different strata. The strata of auriferous dirt differ greatly in thickness and composition; they may be only six inches thick or a hundred feet; they may be of clay, sand, fine or coarse gravel, but usually they are of strong clay, containing gravel and water-worn stones from the size of an egg to that of a bushel basket. Sometimes different layers of auriferous dirt will be found, containing gold of different characters. For instance: there may be a hill two hundred feet deep; at the bottom may be a stratum of blue clay two feet thick, containing much coarse gold; then a stratum of barren gravel three feet thick; then a stratum of yellow clay six inches thick, with fine scale gold; then a stratum of barren sand three feet thick; then a stratum of gray, gravelly clay eighteen inches thick, with coarse shot-gold; and finally, one hundred and ninety feet of red gravelly clay, containing fine gold scattered all through it. It is expected, as a general rule, that the richest dirt will be found next the bed rock; but sometimes rich strata have been found fifteen or twenty feet above the bed rock, and nothing farther down. The diluvial diggings are seldom rich, as compared with the stream placers; they will rarely pay, unless when worked with the sluice.

Ancient and Modern Streams. § 57. The alluvial placers may be divided into ancient and modern. The ancient are those formed by streams which no longer exist, or have found new channels. Two very remarkable examples of the ancient stream placers are found in California; one called the Blue Lead of Sierra county, the other Table Mountain in Tuolumne county. It is supposed that the Blue Lead was once the bed of a large river, about fifty miles eastward of the present position of Sacramento river, and parallel with its course. Table Mountain is a pile of basalt, standing on what was, in the remote past, the bed of a river nearly parallel with the Stanislaus. These ancient and deserted channels are not rare, and are found from a very small to a very large size. They are usually buried at a considerable depth beneath dirt and gravel. Sometimes they are found high above the level of the present streams running near them.

The modern stream diggings may be divided again into the constant and the intermittent—or the never-failing streams, and those which are dry during a portion of the year. The summer and fall are so dry in the interior of California, that only those streams which are very large in the winter run throughout the year, and all the small creeks, brooks, ravines and gullies go dry.

Blake's Classification. § 58. W. P. Blake (*Pacific Railroad Report*, vol. v, p. 217) classifies the placers of California as follows:

1. A coarse, boulder-like drift, the result of great abrasion and powerful currents in a great body of water.
2. A river drift or coarse alluvium, ancient and modern.
3. Alluvial deposits on flats and over broad surfaces, not confined to river channels.
4. Lacustrine deposits, made at the bottoms of former lakes or ponds.

"Lacustrine deposits," says he, "are found in extensive, basin-shaped depressions, in the surface of metamorphic rocks. These depressions have evidently been filled with deep, quiet water, from which thick strata of clay, fine sand and volcanic ashes have been deposited upon the auriferous layer at the bottom. The deposits at Georgetown and Cement Hill are examples of this class. They are more nearly allied to the alluvial deposits of the 'flats' than to either of the other forms under which the auriferous drift appears. The section of the strata at Mameluke Hill shows that the deposition of clay and volcanic ashes from the lake was interrupted, and that a layer of auriferous gravel was spread out over the surface of the clay, after which the former conditions were restored. This alternation, or a series of auriferous and non-auriferous materials stratigraphically deposited, forms a well-marked difference between the lacustrine deposits and those of the flats.

" Great changes have been produced in all of these deposits, of denudation and erosion, during and since the elevation of the region to its present level. The old rivers changed their beds, lakes were drained, and new streams cut their way through great deposits of coarse [diluvial] drift, through lacustrine deposits, and across the ancient river courses. But the action of the denuding streams has not been confined to the superficial deposits, either auriferous or tertiary ; they have eroded great valleys and cañons in the underlying rocks, both of granite, and limestone, and slate ; all are cut through and traversed by long valleys, nearly transverse to the trend of the rocks. These valleys of erosion are on a most magnificent scale, and may be regarded as deep ravines in a formerly unbroken plateau or slope.

" On the forks of the American river, these eroded valleys are from 1,500 to 3,000 feet deep, and the traveler who desires to cross from one bank to the opposite side, must wind in a zigzag line down one side, and in a similar manner up the other, traversing a distance, in most cases, of three miles, while in a direct line it may be but little more than a mile from one bank to the other. All this erosion has taken place since the original deposition of the gold ; and it is probable that the gold of the streams is derived from the original deposits of coarse drift, rather than from the action of the rivers on the veins, although a very considerable quantity of gold must have been liberated from the veins by their action. The great currents or floods which produced the drift, were much more gradual and widespread in their action, and appear to have abraded the whole surface, rather than mere lines or channels."

Position of the Pay-Dirt. ¿ 59. In the auriferous gullies and ravines, the pay-dirt is usually as wide as the current of water during heavy rains. It is customary to speak of "the golden sands of California;" but a person who should believe that the gold is found in pure sand, would be far wrong. Usually, the pay-dirt is a very stiff clay, full of large gravel and stones. The depth of this pay-dirt varies. In a gully where the water is not more than five feet wide in the heaviest rain, the pay-dirt will not usually be more than a foot deep. Between the pay-dirt and the water there is ordinarily some barren dirt, which is thrown away or "stripped off" before washing commences. The barren dirt, as a general rule, contains more sand and less clay than the pay-dirt. Sometimes, when the channel in a small gulch has changed a little, the pay-dirt may be covered with six or eight feet of fertile, loamy soil. In places where the bed-rock of a gully is smooth or steep, the gold usually slips down, and in such spots the pay-dirt will not be so rich as where the bed is level, and where the strata of the rock are nearly perpendicular, and thus present many points to obstruct the passage of the gold. The heavier pieces of gold are ordinarily found near the bed-rock, and in the deepest part of the channel. The diggings on the constant streams or rivers may be divided into "bar" and "bed" diggings. The bars are those banks of gravel and sand which are covered by the river at high water, but are bare at low water; the bed diggings are in those places covered by water at the lowest stage. The bars are usually found at curves of the streams, on the inner side. The richest bars are usually those found near the mouths of cañons; and those spots are richest near where the water has its strongest eddy when the bar is overflowed, and near the limit of low water. The distribution of gold in river beds follows laws similar to those prevalent in regard to gullies. In places where the stream is swift and the bed-rock smooth, little gold is found; and the head of a stretch nearly level is richer than its foot, and the largest pieces are found in the deepest parts of the channel, and so forth.

Geological Character of the Coast. ¿ 60. The geology of this coast has never been thoroughly studied. The great bulk of the Sierra Nevada and the Cascade mountains is composed of granite, covered in places by trap, trachyte, basalt, serpentine, limestone, slate, lava, quartz and diluvium.

The Coast Range is a sandstone, with occasional trap and granites. The valley between the Sierra Nevada and the Coast Range, extending from Puget Sound to Tejon Pass— for it was originally but one valley, though now separated by the Siskiyou, Umpqua and Calapooya mountains—is a mass of diluvium, from 200 to 500 feet deep, lying on tertiary sandstone. The country east of the Sierra Nevada is mostly covered with trap, other eruptive rocks of various classes, and granite.

Volcanic Mountains. ? 61. Many of the mountains are of volcanic origin, and all the highest peaks were or are volcanoes, including Mts. Baker, Rainier, St. Helens and Adams, in Washington Territory, Jefferson, Hood, McLaughlin and the Three Sisters, in Oregon, and Shasta, Lassen's Peak, the Downieville Butte and Castle Peak, in California. Mt. Baker is now an active volcano ; Mt. Helens emits a constant stream of steam-like smoke, and Shasta sends hot and sulphurous vapors from her summit.

Beds of lava which have been thrown out by volcanoes, are not rare in the mining districts, and they occupy an important place in the geology of the gold region. In several places they have covered up the beds and channels of ancient streams, thus covering rich deposits of gold ; and subsequently they have been left, by the washing away of the adjacent ground, standing as high, narrow ridges, with flat tops and steep sides. The most singular and remarkable " table mountains," as they are called, are—one in Tuolumne county, thirty-five miles long ; the other in Butte county, about ten miles long.

Hot Springs. ? 62. It has been observed that where warm and hot mineral springs are found over an extensive district of country, there are usually valuable minerals in the soil. Such springs are rarely found in fertile, flat or gently undulating countries, where the rocks next the surface are stratified limestone or old red sandstone ; but in rugged, mountainous places, where the rocks are granite or volcanic in character. There is probably no part of the world richer in hot and mineral springs than California and western Utah. The Coast mountains, from latitude forty degrees southward to thirty-two degrees, are full of warm springs, most of them with a good portion of sulphur in their waters. On the western slope of the Sierra Nevada, hot springs are rare, but on the eastern side most abundant ; and, indeed, there are places where, in the midst of springs and brooks, it is difficult to find water free from sulphur and iron.

CHAPTER V.

THE MINING DISTRICTS.

Orography of the Coast. ¿ 63. It may be said, in general terms, that the whole Pacific coast of North America is rich in gold and silver, from the Isthmus of Panama to the Russian Possessions.

In considering the mining districts of California and the adjacent States and territories, we must remember the general conformation of the country. The back-bone of the coast is the range of mountains known as the Sierra Nevada in California, and the Cascade mountains in Oregon, Washington and British Columbia. This range is from twenty to fifty miles wide, from five to ten thousand feet high, about a hundred miles from the coast and parallel with it. It is metalliferous through nearly its whole length, from thirty-four degrees to fifty-five degrees, and its southern termination is connected with the Coast range which runs down through the peninsula of Lower California, and is also metalliferous.

Metals of the Sierra Nevada. ¿ 64. The richest auriferous region on the coast is found on the western slope of the Sierra Nevada, between latitudes thirty-seven and forty degrees. On the eastern side of the summit, between those latitudes, a few rich placers have been found, some very rich quartz veins within one small district, and a large extent of country rich in silver. South of thirty-seven degrees, there are a few scattering placers, and auriferous quartz veins west of the summit; none east of it. From forty degrees to forty-three degrees, there are numerous placers and auriferous quartz lodes west of the summit; on the west side, from forty-three degrees northward, and on the east side, from forty degrees to forty-six degrees, there are no paying placers or veins of gold or silver. From forty-six degrees to fifty-eight degrees on the eastern side, there are paying gold placers. The gold found east of the summit contains a greater proportion of silver than that found west; this remark applies equally to gold dug in latitude fifty-two degrees, and in thirty-eight degrees.

General List of Districts. § 65. The Sacramento mines, or the auriferous region of the Sacramento basin, about half way up on the western slope of the Sierra Nevada, is forty miles wide and two hundred miles long, lying between latitudes thirty-seven and forty degrees, with a direction north-north-west and south-south-east longitudinally.

The next gold bearing district is in the basin of the Shasta and Klamath rivers, between forty deg. thirty min. and forty-three deg. thirty min., extending about two hundred miles along the coast and one hundred miles inland. Northward of forty-three deg. thirty min. few mines are found west of the Cascades.

The third auriferous district in importance is that of British Columbia, on the eastern side of the Cascade mountains, between fifty and fifty-five degrees, and about fifty miles wide.

The basin of Rogue river, in southern Oregon, contains rich and extensive placers.

The next district in importance is that of the valley of the Upper Columbia, in Washington territory, east of the Cascades.

There are gold diggings also in the valleys of the Walker, Carson, Kern, White, San Gabriel, Colorado and Gila rivers, at Mono Lake and in Arizona.

The richest argentiferous district is that of Washoe, on the eastern side of the Sierra Nevada, in latitude thirty-nine deg. twenty min. It is about ten miles square.

Ninety miles southward and fifteen miles northward from the eastern point of Mono Lake, is the Esmeralda district, about five miles square. This district is supposed to be within the limits of Calaveras county, California.

Two hundred miles further southward, and eastward from Owen's Lake, is the Coso district, about ten miles square.

Arizona contains gold, silver and copper; Lower California has silver and copper.

The only quicksilver mines on the coast, are found in the

vicinity of New Almaden, Santa Clara county; at New Idria, in Fresno county; at Mt. St. Helena, in Napa county, and the Geyser mountains, in Sonoma county.

Climates. ₴ 66. The climates of these various mining districts differ greatly. The quicksilver mines lie at a low level —about a thousand feet above the sea; are within forty miles of the coast, and have a mild, equable climate. At New Almaden, ice never forms more than two inches, and then thaws within a day or two; snow never lies on the ground more than a day, and never falls more than two or three inches deep; the thermometer is never below the freezing point twenty-four consecutive hours; there are three hundred clear days in the year, and in summer the heat is never oppressive. Sonoma and Napa counties and New Idria have a similar climate. About twenty inches of rain fall between November and May; none during the remainder of the year. The nights are cold all along the coast, in summer as well as winter.

The mining district on the western slope of the Sierra Nevada has hot summers and cold winters. Snow lies on the ground for weeks at a time, and ice forms nearly a foot thick. The severity of the winter depends upon the altitude. At Forest City, in Sierra county, about six thousand feet above the sea, the snow falls sometimes ten feet deep, and lies on the ground from December till March. The people often use snow shoes in traveling. The summers are dry, the winters wet, and the amount of rain increases with the altitude.

About the mouth of the Klamath river and in Rogue river valley, rain is not rare in summer. Near the ocean, the climate is mild and equable in temperature; but in the vicinity of Yreka, the winters are very cold.

The climate of British Columbia is very cold in winter. The summers are hot at mid-day. The same remark may be applied to the valley of the Upper Columbia. Rain is rare in summer.

In western Utah, the winters are cold, the springs windy and gusty, the summer days warm. Rain is rare between May and November.

In Arizona, the heat is great in the summer; the winters are warm and mild.

Gold is found in most of the counties of California, including all that reach the eastern border of the State.

Sacramento District. § 67. The principal mining district, which may be called the "Sacramento district," because it is in the Sacramento basin, lies on the western slope of the Sierra Nevada mountains, between latitudes thirty-seven degrees and forty degrees north, and between one and six thousand feet above the level of the sea. It is one hundred and seventy-five miles long by thirty wide, to speak in general terms, and its longitudinal direction is north-north-west by south-south-east. It includes the greater portions of the counties of Mariposa, Tuolumne, Calaveras, Amador, El Dorado, Placer, Nevada, Sierra, Yuba, Butte and Plumas; a very mountainous, rugged country—cut up into hills and cañons by numerous streams running westward from the summit of the Sierra, and emptying into the Sacramento and San Joaquin rivers. These streams, commencing at the north, are the Feather, which has the Yuba and Bear rivers for tributaries; the American; the Mokelumne, which has the Cosumnes for a tributary; the Calaveras, the Stanislaus, the Tuolumne, the Merced, the Mariposa, and the Chowchilla. These streams are swift mountain torrents, each about seventy-five miles long on an average, running through deep rocky gorges and perpendicular cañons. A level valley, half a mile wide and five miles long, is a great rarity along these streams, which, though large, have not been able to win much level land from the mountains. The country is so rough, that in '49 few of the mining camps were accessible with wagons, and notwithstanding the extensive trade of the mines, and the expenditure of large amounts on the roads, pack mules are still extensively used as the means of conveying freight. Part of this mining district is covered with a dense forest of evergreen trees; part of it has scattered trees, and part of it is bare, or covered with thick bushes. The limits of this district are not clearly defined; gold is found in small quantities on all sides of it, and many places, within its general limits, have no gold.

The Sacramento district has been divided by common usage, into the "Northern" and "Southern" mines. The "Northern mines" lie north of the Mokelumne river, comprise the counties of Plumas, Butte, Yuba, Sierra, Nevada, Placer, El Dorado and Amador, and get their supplies through Sacramento. The "Southern mines" lie south of the Mokelumne river, comprise the counties of Calaveras, Tuolumne, Mariposa and Fresno, and get most of their supplies through Stockton.

Nearly every county includes all kinds of diggings; and every little camp has something different from other camps.

Plumas County. ? 68. Plumas county has some rich mining land in her south-western corner, but it is very high above the sea. Most of the diggings are deep. Quincy, the county seat, is one hundred and forty-five miles from Sacramento.

Sierra County. ? 69. Sierra is one of the smallest but richest mining counties in the State. Little, if any of it, is within four thousand feet of the sea level. Its diggings are chiefly in hills, and very deep; and they will continue to yield largely for many years. The gold is coarse. The country is well supplied with water. A remarkable part of the mining ground of this county is called the "Blue Lead," of which Mr. Charles S. Capp, says:

"This is not one of the many petty leads, an inch or two in breadth and thickness, which, after being traced a few hundred feet, end as suddenly and mysteriously as they commence; but it is, evidently, the bed of some ancient river. It is often hundreds of feet in width, and extends for miles and miles, a thousand feet below the summits of high mountains, and entirely through them. Now it crops out where the deep channels of some of the rivers and ravines of the present day have cut it asunder; and then, hidden beneath the rocks and strata above it, it only emerges again, miles and miles away. Wherever its continuity has been destroyed, the river or gulch which has washed a portion of it away was found to be immensely rich, for some distance below, and the materials of which the lead is composed are found with the gold in the bed of the stream. It is evidently the bed of some ancient stream, because it is walled in by steep banks of hard bed-rock, precisely like the banks of rivers and ravines in which water now runs, and because it is composed of clay which is evidently a sedimentary deposit, and of pebbles and black and white quartz, which could only be rounded and polished as they are by the long continued action of swiftly running water. The bed-rock in the bottom of this lead is worn into long, smooth channels, and also has its roughnesses and crevices like other river beds. The lighter and poorer qualities of gold are found nearest to its edges, while the heavier and finer portions have found their way to the deeper places, near the center. Trees and pieces of wood, more or less petrified and changed in their nature, which once floated in its waters, are also everywhere encountered throughout this stratum.

"The clay and finer gravel in which these pebbles and boulders are found to be tightly packed, is of a light blue color, which gives the name to the lead. Much of this clay is remarkably fine and free from coarser particles, and is smooth and unctuous to the touch. It is said to be strongly impregnated with arsenic, as was shown by chemical analysis, and contains large quantities of iron and sulphur in solution, for pyrites and sulphurets of iron are deposited in shining metallic crystals in every vacant crevice. Fine gold is found among this clay, and the heavier particles beneath it, upon the bed-rock. This stratum varies in thickness from eighteen inches to eight or ten feet, while the whole lead varies in width from a hundred and fifty to five hundred feet.

"The same lead has been found at Sebastopol, four miles above Monte Cristo, and also higher up among the mountains. It appears at Monte Cristo, which is four miles above the high-lying Downieville, and over three thousand feet above it, and at Chapparal Hill, on the side of a deep ravine; then at the City of Six, which is also on very high land, about four miles from Downieville, across the North Yuba. It is next found at Forest City, on both sides of a creek, and is there traced directly through the mountain to Alleghany town and Smith's Flat, on the opposite side. There it is again cut in twain by a deep ravine. It crops out on the other side at Chip's Flat, where it has been followed by tunnels passing completely through the mountain to Centreville and Minnesota on the other side. Here it is obliterated by the Middle Fork of the Yuba, but is believed to be again found at Snow Point, on the opposite side of the river; and again at Zion Hill, several miles beyond. There is no reason for doubting that after thus reaching over twenty miles, it still extends further. Hundreds of tunnels have been run in search of it. Where the line it follows was adhered to, they have always found it, and have been well rewarded for their labor. Millions of dollars have been taken from this lead, and its richness, even in portions longest worked, is yet undiminished. These tunnels have cost from $20,000 to $100,000 each, and interests in the claims they enter sell readily at from $1,000 to $20,000, in proportion to the amount of ground within them remaining untouched, and the facilities which exist for working it. Many of these claims will yet afford from five to ten or more years' profitable labor to their owners, before the lead itself within them is exhausted. As in some of them quartz veins and poorer paying gravel have been found, many of them may be valuable to work from the top down, as hydraulic claims."

The theory that this blue lead was once the bed of an ancient stream, is generally accepted by those familiar with it. Another evidence to support the theory is, that in many places the flattish stones in the lead lie at a peculiar inclination, and all in the same direction, as stones do in a stream of water. This theory, however, does not find universal belief. Mr. B. P. Avery wrote thus of it for the San Juan *Press*, in 1859:

"Everybody in California has heard or read of the famous 'blue lead,' which all miners, who delve for gold far up in the mountains, hope to find, and think themselves lucky when they have found it, and which they pronounce to be the channel of an ancient river. This lead is always found resting on or near the bed-rock, beneath diluvial strata of different colors, such as shades of red, yellow and gray, and is itself more of a deep slate color than blue. It is generally richer in auriferous particles than the gravel lying above it, and forms the productive drift diggings for which the vicinity of Forest City is noted, as well as those of many other localities. The theory of its origin alluded to above, is predicated upon these facts and assumptions: that it has been traced, in a continuous line, at a certain altitude, through several counties, from ridge to ridge, at a right angle to present water courses, across cañons thousands of feet deep; that the stratification of the lead is uniform, and different from that of adjoining deposits; that tree trunks, both in the ligneous and petrified state, are found lying in it, as though borne there by freshets; and the gold found in it is everywhere of the same character as to appearance and quality. This crude theory conforms to the general one, which is popularly employed to account for the extensive alluvial deposits constituting our placer diggings. It is remarkable that the majority of our miners, who are commonly men of intelligence and practical knowledge in their pursuit, should have discarded entirely, if they ever entertained, when speculating upon the origin of our gold fields, the more rational theory of marine influence, for one of purely local causes.

"They overlook all the facts which go to prove a total submergence of this coast at some remote period, and settle down upon the narrow idea that the immense gravel beds which contain so large a portion of our mineral wealth, and which extend at least four hundred miles north and south, having an average breadth of probably not less than sixty miles, were deposited by rivers, which anciently ran here, and changed their channels from time to time, until they paved the whole country with cobble stones. These deposits have been cut through by modern streams, running a different course, and hence the present cañons and ridges. Of the ancient rivers, the one that deposited the blue lead has alone left distinctive marks of its course. Now, unfortunately for the plausibility of this theory, the blue lead is found all the way from the summit of the Sierra Nevada to the foot hills. Instead of being confined to a certain altitude, and a certain line, it exists in every altitude, on the main ridges as well as on spurs of them, and even on isolated peaks. Its color is owing to the presence of sulphuret of iron in solution, without which the gravel would not be any different from that lying above, except that the boulders and large stones would be found in it as they are always found at the bottom of every gravel deposit. Wherever sulphurous acid or sulphuret of iron is found, there the so-called blue lead will be discovered, just as certainly as red earth and gravel will be found where the oxide of iron is present as a coloring agent. It is found at a great elevation in Sierra county, and at a low one in Nevada and Yuba. It has been struck at San Juan, and at points thirty or forty miles above it, leads of other colors intervening."

The quartz claim of the Sierra Buttes Quartz Mining Company in Sierra county, is one of the most valuable in the State. The claim derives its name from the Sierra Butte or Buttes, a mountain about 9,000 feet high, with three peaks, twelve miles north-eastward from Downieville. On the south-western slope of this mountain, about 6,000 feet above the level of the sea, and twelve miles from the county seat, are the claim and mills of the company. The company is composed of Messrs. Reis Bros., who have a half interest, Wm. H. Ladd, of Downieville, who has a fourth, and E. L. Barnes, of Downieville, and R. E. Brewster, of San Francisco, who have each an eighth.

The lead was discovered in 1851, but little work was done previous to 1853. The claim is 1,600 feet long, and includes several lodes, only two of which have been worked. Of these, the larger one, called the Cliff ledge, is from four to twenty-five feet wide; the smaller one, called the Aerial ledge, is from two to four feet wide. Little work has been done on the latter. Both these lodes have a direction nearly north-west and southeast, and both dip to the east, at an angle of forty-five degrees, though at some distance from the surface, the dip seems to approach a little nearer to the horizontal direction. The quartz is bluish white in color, and hard, but much of it crumbles into sand on exposure to the air.

The paying rock is, in some places, not more than two feet thick; in others, as much as eighteen; on an average, six feet. The richest rock is found in a streak near the foot-wall; and there are rich spots scattered around, irregularly, in other parts of the lode. The claim of the company is on a steep mountain-side, and the rock is taken out through tunnels. The lowest point where the lode has been struck, is 425 feet below the outcropping of the Cliff lode. The vein is about the same in size, and the rock the same in quality and in richness there, as at the surface. The rich streak along the foot-wall is from four to thirty inches thick, and pays at the rate of from $50 to $5,000 per ton. This rich quartz, however, is only obtained in small quantities at a time, and it is always crushed with the other poorer rock. The average yield is $18 to the ton. The gold is from 865 to 870 fine, and is, therefore, worth from $17 88 to $17 98 per ounce. The claim has a high reputation for the great amount of paying rock.

The gold is extracted in two mills, owned by the company. These mills are in a ravine about 1,000 feet below the out-crop of the Cliff lode. One has twelve revolving stamps, the other eight square stamps, which last are soon to be taken out and replaced by revolving stamps. These mills run day and night, and crush from twenty-eight to thirty tons in twenty-four hours. The quartz is carried down from the tunnels to the mills on tramways, in cars. A loaded car going down, pulls an empty one up. The quartz is all loosened by blasting. · Forty men are employed by the company; twenty-five to quarry the rock, five in the mill, to look after the stamping and amalgamating, a carpenter, blacksmith, and eight others to get out timber, transport quartz, and so forth.

The amalgamation commences in the battery, where about two-thirds of the gold are caught. From the battery the gold is carried over an amalgamating copper plate, then over blankets, then through a little sluice with transverse riffle-bars, and then through a sluice paved with cobble stones. The mills run continuously night and day, week days and Sundays, stopping only to clean up. A run occupies from thirty to forty days. Four pounds and a half avoirdupois of quicksilver are put into the battery daily. All the amalgam is retorted, in masses worth about $6,000 each. The average cost of quarrying, crushing and amalgamating a ton of auriferous quartz, is six dollars; and as the yield is eighteen dollars, there is a profit of twelve dollars per ton. At twenty-eight tons per day, about 10,000 tons are crushed in a year, giving $120,000 profit, and a gross yield of $180,000. One of the mills only got into proper operation in April, 1860, so they have not yet had a full year of experience with both of them. The dividends of 1860 amounted to $83,000, exclusive of $9,000 spent on the new mill.

The wages of the men hired are from fifty to seventy-five dollars per month—average, sixty dollars and boarding. Until of late, the tailings were allowed to run off, but now they are saved, and the company has made a contract to have them worked with three arastras, on shares. There is not much pyrites in the quartz.

The mills are driven by water, supplied by two flumes, one 1,000 feet long, the other 5,100. All the buildings are erected on a hill-side, which rises at an angle of about thirty-five degrees, so that all the provisions have to be brought on mules, and such firewood as cannot be floated down the flumes, must be obtained by the same conveyance. Heavy timbers and castings for the mills are let down the mountain-side by a block and tackle.

A correspondent of the Sacramento *Union* wrote thus about Howland Flat, in this county, in December, 1860:

"Howland Flat has eight companies in successful operation. Three of them have steam engines for raising dirt. The Union Company is, I presume, one of the most extensive mining companies in Northern Sierra. They have a large engine, by which they draw their dirt up an incline two hundred and forty feet long, and at the same time pump the water out of their diggings. Their main tunnel runs back under the far-famed Table Mountain near two thousand feet, where they get their rich pay. As is generally the case, those who open heavy claims, scarcely ever reap the benefits.

"The first company was known as the Bright Star Company, expended forty or fifty thousand dollars, and then failed, and the present company took charge, and are doing remarkably well. Claims are valued very high in this company. The Mountaineer Company has also an engine, by which they hoist dirt through a shaft one hundred and thirty-seven feet deep. They just got under way last fall, and their claims have been paying large—an ounce to the drifter. The Minnesota Company, one of the oldest on the flat, has an engine, and, as it has always done, pays well. A good 'pile' for more than one has been taken out of those claims, and still there are more fortunes in there yet. The Down-East is another old company, and pays regularly from an ounce to twenty dollars a day to the drifter. The Shirley Company is another of those good paying claims—never fails to reward the laborer. The Golden Age is an old claim, and is paying splendidly. The St. Louis Company was once abandoned, but changed hands, and is now paying handsomely. The Golden Era Company had just commenced hoisting dirt the day I visited them, by a process they call a 'water-balance.' They hoist a car-load at a time. There is a large chain which runs over a wheel, and at each end of this chain is attached a water tank, and, when filled, will just balance a loaded car, which sits immediately over the empty water tank. When they wish to hoist a car-load of dirt, they fill the tank with water, which is on top, and it starts down the shaft, leaving with it the empty car, and at the same time bringing up the empty tank and car-load of dirt. When this tank, full of water, strikes the bottom, the water flows out preparatory to receiving the loaded car. They have a large tank in 'whim' which supplies these little tanks. It takes two inches of water per day, which costs the company sixty cents per day to run the machine, is all the expense they are at, and they can hoist as much dirt as any of the engines. The beauty of this machine is the cheapness and simplicity of it. It is singular to me there are not more of them in operation in California, as they would do away with expensive engines."

Downieville, the county seat of Sierra county, is one hundred and ten miles from Sacramento.

Butte County. § 70. The diggings in Butte county are near the level of the Sacramento Valley. Oroville, the county seat, on the bank of Feather river, is seventy-five miles from Sacramento, in a due northward direction. Most of the mining in this county, until within a couple of years past, was done in the bars and beds of Feather river and its tributaries; but since 1858, these have been pretty well exhausted, and now recourse has been had to hills and flats. Northward from Oroville is Table Mountain, a hill of basalt, ten miles long, overlying a bed of rich auriferous clay and gravel. The bed of Feather river was once extremely rich, and it paid for some of the grandest mining enterprises ever undertaken in the State. The chief of these were the works of the Cape Claim and the Union Cape Claim companies, in 1857 and 1858. These companies took up the waters of Feather river, in a flume several miles long, and cleaned out its bed. The Cape Claim flume was three-quarters of a mile long and twenty feet wide. Two hundred and fifty men were employed. On the fifth of November, 1857, the mining operations of the company having been closed for that year, the treasurer reported that the expenditures of the season, (the claim being in the river bed, could be worked only during the summer) had been $176,985, and the receipts $251,426, showing a profit of $74,441. The next summer the company continued its enterprise; expended $160,000, received $115,000, and so lost $45,000. The distance from Sacramento to Oroville, the county seat of Butte county, is seventy-five miles, by the stage road.

Nevada County. § 71. Nevada county is the first mining county in the State. It is the largest in population, probably the richest in gold, and it has done more for the advancement of mining than any other district. The tom, the sluice and the under-current sluice were invented there, and the ditch and hydraulic process were there first applied to mining purposes; and there, also, quartz mining was first undertaken on an extensive scale, and carried into successful operation. The chief mining towns of the county are Nevada, Grass Valley, North San Juan and French Corral. Grass Valley is noted for its quartz mines; the two last named places, for their extensive hydraulic diggings. Sweetland, a small town, is famous for its tail sluices. The Allison mine is, next to the Fremont mines in Mariposa, the richest quartz mine in the State. It is supposed to yield about $40,000 per month. It was discovered in 1856, by some men engaged in placer mining. They found that the dirt on one side of the creek bank, where they were sluicing, was very rich, and they followed it up till they came to the quartz vein from which the gold had come; and on testing the quartz, they found it exceedingly rich. It is said that the vein has yielded $1,500,000, and that most of the rock is worth $300 per ton, on an average. The main shaft in the vein is one hundred and fifty feet deep.

"Nearly all the mining now carried on around the town of Nevada," says Mr. Capp, "is either in hydraulic claims, supplied with water from the flumes and ditches, or else consists in washing, in the bed of the creek, the tailings that annually enrich it. In some instances, below hydraulic claims, long sluices are arranged so as to catch all the tailings and water as they are discharged, and give the fine gold that has escaped another chance to settle. This arrangement is also an advantage to the parties above, who thus do not need to make their sluices so long as would otherwise be necessary, and are at little or no expense in keeping them free of stones and tailings, the force of the current being preserved, and the other parties attending to this business. I heard of one instance where a long sluice, erected for this purpose, and receiving no other attention whatever, is cleaned up by one man, once or twice a month, and yields usually a dividend of about seventy-five dollars. In other cases, two or three men, from such a sluice, by working constantly to keep it clear, make fair wages.

"Most of the hills about Nevada were worked in the early times of '50 and '51, by shafts or wells, from fifteen to sixty feet in depth, from which drifts branched off in all directions, following up the many little leads and extracting the richest of the pay-dirt. Now, such claims are sufficiently rich to pay well when worked in the hydraulic fashion, which is done wherever water is to be had."

The Eureka Lake Ditch, owned by a company of the same name, is the most extensive enterprise of the kind in the State. The main ditch is seventy-five miles long, and it has one hundred and ninety miles of branches, making two hundred and sixty-five miles, which have cost $900,000. The daily sale of water is six thousand inches, at sixteen cents per inch, which makes a weekly revenue of six thousand dollars. The water is obtained chiefly from about two dozen little lakes, very high up in the Sierra Nevada, and large artificial reservoirs have been built to hold, until late in the fall, the water collected in the winter and spring.

It is said that $8,000,000 has been taken from a hill a mile long and half a mile wide, at Coyoteville, near Nevada. One of the most important districts in the State, for hydraulic mining, is the French Corral ridge, a hill about thirty miles long, a mile wide and two hundred feet high, in the north-western part of Nevada county. This ridge runs from the mining camp of French Corral to North San Juan. The town of Nevada, the county seat, is sixty-nine miles from Sacramento.

Yuba County. § 72. Yuba county contains rich mining ground in its north-eastern portion, though its south-western part is in the flat land of the Sacramento Valley. The most important mining towns of the county are Camptonville and Timbuctoo. Marysville, the county seat, is forty-four miles from Sacramento, by the stage road.

Mr. B. P. Avery, of the San Juan *Press*, wrote thus about Timbuctoo in December, 1859:

"The diggings of Timbuctoo mostly consist of a gently-rounded gravel hill, about one mile long and perhaps not over fifteen hundred feet wide, with an altitude above its base of about three hundred feet. This hill lies in a north-east by a south-west direction, between a large ravine, where the town is, on one side, and the Yuba river on the other. It is partly bisected by a small gulch. The gravel, which is very coarse and easily washed, but for intermediate strata of 'pipe clay' and indurated sand, is not less than two hundred feet deep, has an inclination towards the river, and rests upon trap rock, which is highest next the stream. The first deposition upon the rock is a stratum of cemented gravel, deriving its color and hardness from iron sulphurets, and very rich in gold worth eighteen dollars and fifty cents per ounce. Overlaying this is a thick stratum of indurated clay or sand—for it varies in character—containing some gold, but indissoluble except after being cut up and long exposed to air and water.

"The blue gravel beneath has to be drifted out, and is washed several times in succession, through sluices, the tailings being saved by means of dams and allowed to 'slack.' Above the clay stratum the gravel is yellow and gray, very coarse, crumbling down readily beneath the action of one hundred inches of water, forced against it from hose and pipe. This top-gravel, as it is called, is from one hundred to one hundred and forty feet thick, and is washed off entirely by the hydraulic process. The auriferous particles are not distributed through it uniformly, but are found most plentiful in 'streaks,' or thin strata following the plane of the bank. Some of these strata, lying fifty and one hundred feet above the bed-rock, have been drifted, and paid from five to ten dollars a day per man. The gold in the top-gravel is remarkably fine in quality, commanding from nineteen dollars and twenty cents to nineteen dollars and fifty cents per ounce from buyers, and assaying as high as 987.

"The diggings were discovered about eight years ago, by parties following up ravines from the river. One of these ravines contained a great deal of earth that yielded one dollar to the bucket. The blue gravel in the hill was much better. As an instance of this, we are told that ten dollars had been obtained from a single prospect; and three claims, worked by Antoine, Burgoyne and Boyd, yielded, in something over two years, about $100,000; several parties buying in successively at round prices, and going home with $5,000 or $6,000 not long afterwards. But drifting out bottom dirt was found to be less profitable, on the whole, than sluicing off the top.

"The hydraulic process, in its crudest form, with small sluices and small streams of water, was first employed between four and five years ago. It is now the universal mode of work, and is conducted with elaborate system; galvanized iron pipes leading the water into the claims, where it is distributed through rubber and canvas hose; substantial sluices carrying off the washings, and more extensive ones receiving the united 'tailings' of several sets of claims. As the bed-rock lies so low, there is no necessity for tunnels, except in two or three instances. The only difficulty the miners labor under is want of sufficient fall or grade for their sluices. The distance from the diggings to the river is nearly a mile, and the utmost grade that can be had for this distance, will not exceed eight inches to every twelve feet length of sluice. The big ravine through which the tailings run, is filled with them to a depth of sixty feet. Along this ravine are several large, tall flumes.

"Water was first brought into the diggings of Timbuctoo and vicinity, by ditches from Deer Creek, twelve miles long. Three of these, one constructed by Mr. Bovyer, noted for his enterprise and skill in such works, are now consolidated and owned by the Tri-Union Company. The water which they supply, amounting, during the wet season, to 4,000 or 5,000 inches, is conveyed across the gap lying between Smartsville and Sucker Flat, through two flumes 2,400 feet long; the largest one, which appears to be sixty or eighty feet high over the lowest part of the gap, being an elegant and substantial structure. The Excelsior company have another ditch, which is twenty-eight miles long, draws its water from the South Yuba, has a capacity of about 4,000 inches, miners' measure, and cost in the neighborhood of $250,000. It will deliver 3,000 inches, as water is measured at Timbuctoo, that is—under a pressure of ten inches—equal to perhaps 4,000 inches as measured at North San Juan, under six inches pressure. The Excelsior Company is now building, at Empire Ranch, two miles from the diggings, a reservoir which will have a depth at the dam of not less than fifty feet, and a superficial area of at least seventy-five acres. The dam is about one hundred and fifty feet wide at the base, and pierced by a very substantial arched conduit of stone and wood work. The water will be admitted through this, when wanted, by means of a heavy iron gate, raised by a screw—a contrivance of great power and steadiness. The ditches enumerated are now supplying the diggings with about 8,000 inches of water, at twenty-five cents an inch, measured liberally, as above stated. Of this quantity, the largest portion is used at Timbuctoo, where forty companies and nearly three hundred men are at work, though not more than half that number of companies are washing. Each company uses from 400 to 600 inches of water; and the day we were at Timbuctoo, the total water sales on the hill reached 6,000 inches, at a cost of $1,500. Of course, this enormous supply dwindles to a very small one as the dry season approaches, and many companies have to lie by or go to drifting. Winter is the true harvest time of both ditch men and miners. During the past winter, we are told, the weekly receipts of Bovyer's ditch averaged $3,000.

"Although the miners use such large heads of water, they do not direct it all against the bank. Only one hundred inches flow through the pipe, the remainder being allowed to fall over the edge of the claims; much less being needed to get the earth down, than to run it off, owing to the looseness of the upper gravel, and great size of the stones. In some claims, however, the earth is more compact, and to facilitate washing, the bottom is drifted out, leaving pillars standing, which are piped away, and then down comes the mass above."

Placer County. ? 73. The principal mining towns of Placer county are Auburn, (the county seat) Iowa Hill, Yankee Jim's, Todd's Valley, Wisconsin Hill, Michigan Bluff and Dutch Flat.

"The stratification of the rocks in the neighborhood of Sarahsville," says Mr. Capp, from whom I shall have occasion to quote frequently, "is peculiar, there being two distinct layers of pay-dirt. The cement of which the top of the ridge is composed, is sometimes three hundred feet in thickness, but is much thinner where the bed-rock rises. Directly under this cement is a stratum of pay-dirt, called the upper lead. It is a reddish or grayish gravel, and from two to six feet in thickness. Below this is a stratum of blue gravel, of variable thickness. It is thinnest where the bed-rock is highest. In the deepest places it is one hundred and thirty feet in thickness. Under this third stratum is the lower lead of pay-dirt, which lies immediately upon the bed-rock, and is from six to eighteen inches in thickness. The upper layer of cement is whitish and bluish, and where exposed to the air and moisture on the tops and sides of the hills, it is decomposed, and forms the red gravel which is a distinguishing feature of nearly all gold mining districts.

"The first stratum of pay-dirt, a reddish gravel, contains coarse boulders at the bottom, but is composed of much finer particles towards the top, and has evidently settled, as a sedimentary deposit, from running water. The stratum of gravel below this contains more or less gold, being "spotted" as the miners call it, portions having yielded forty cents to the pan, while in others the color of gold is barely perceptible. At some future time this stratum may pay to wash in the hydraulic fashion. The lower stratum of pay-dirt is a fine sediment, composed of quartz and the bed-rock of the neighborhood, which very much resembles a greenish soap-stone, ground up together and containing pebbles and boulders of quartz. The gold in the top-lead is in small, bright scales; that in the lower lead is rounder and heavier, and the surface of some of it is coated and discolored. To work these two leads, it is necessary to run separate tunnels, at from fifty to a hundred feet apart, one above the other. By tracing the slope of the bed-rock on which the lower lead of pay-dirt is distributed, it has been found that the tunnels near Snowburg (which is merely separated by a point of rocks from Sarahsville, and lies to the eastward of it) are running into the ridge along the course of an old river or ravine. The distance across the bottom of the upper stratum of cement, is about two thousand feet, and from its edges the bed-rock dips rapidly both ways towards the center, thus forming the banks of the old channel. The bottom lead of pay-dirt follows down the steep sides of the bed-rock, and is supposed to be thickest and richest in the center claim, where, however, the lower tunnel has not yet been completed. This channel appears to have crossed the dividing ridge between Shirt Tail cañon and the middle fork of the American river. A section of it is partially exposed, and the pay-dirt has cropped out at this point, near the mouth of Volcano cañon.

"Forest Hill is a small mining town, about three miles from Yankee Jim's. It is built on the steep side of the ridge along the west side of the Middle Fork of the American river. The mining operations are carried on almost entirely by tunnels, entering the mountain from the eastern side.

"The pay-dirt in these claims is found on two level surfaces of the bed-rock. The outer rim, through which the tunnels run, is from five hundred to eight hundred feet in thickness. At this depth its inner side is reached, which slopes inwards and downwards towards the center of the mountain, until it reaches the first level, which is usually in the neighborhood of four hundred feet in width. The excavations are carried directly across this flat, and at the furthest side the bed-rock again suddenly 'pitches,' or slants downwards, until another flat is reached, which is about one hundred feet in width. This second flat appears like the ordinary bed or channel of a river, and is bounded on the other side by another sudden rise of the bed-rock, which apparently was the bank or wall that confined the stream. Thus a section of the mountain at this point, it would seem, would exhibit a bed of some old river, almost one hundred feet in width, worn deeply into the bed-rock. At the side where the tunnels now enter, and a few feet above it, a large flat or bottom four hundred feet in width would appear to have been formed, which was also bounded by a steep bank, that at present is the outer or rim-rock.

"The first deposit of pay-dirt that is encountered, is that on what is supposed to be the flat along the true bed of the stream. It is what the miners call 'a white sediment,' filled with quartz pebbles and boulders, but containing little gravel. That found beyond the pitch in the bed-rock on the second level, which is supposed to be the old bed of a stream, is called the main or back lead, and is a finer sedimentary deposit, which the miners call 'pipe clay.' These deposits, I was informed, are about three hundred feet below the top of the mountain. This formation is said to extend for about two miles and a half under the mountain, and nearly following the course of the ridge to where an outlet appears to have existed at Todd's Valley. There, the pay-dirt being more scattered, is supposed to have formed the surface diggings in that vicinity. Not a tunnel run into the mountain on this line has failed to reach these leads, which therefore are evidently continuous."

Two-thirds of the present gold yield of Placer county are obtained from hydraulic claims.

El Dorado County. § 74. El Dorado county, for a long time called the "Empire County," because it cast the largest vote in the State, is now next to Nevada in population among the mining counties. The principal towns are Placerville, Coloma, Georgetown, El Dorado, Spanish Flat and Diamond Springs. Placerville, the county seat, is fifty-one miles from Sacramento in an eastward direction. El Dorado is the oldest placer mining county in the State, Marshall's discovery having been made within its limits.

"The surface and creek diggings in the immediate vicinity of Placerville," said Mr. Capp, writing in 1857, "have been pretty thoroughly worked. The majority of the miners who derive their supplies at this point, are scattered for a number of miles around; and it is this trade and the business that has to be done at the county seat, perhaps more than the mines close by, which principally supports the town. The wages made by miners near Placerville, are as low as in any other portion of the mining region. Most of the mining now going on near the town is washing over old tailings, and tunneling into the hills. Though many such operations have been very profitable, I was shown large numbers of places where little or no return had been received for the labor expended. These operations have been numerous in the Coon Hollow ridge, along the Cedar ravine, at Smith's Flat, and at Negro Hill, two or three miles distant. At the two last named places such operations pay well. The Basset claim, on Ringgold Hill, has been one of the most remunerative in the State. Its richness was discovered by sinking a perpendicular shaft, from which a large amount of gold was taken, but water finally drove the miners out. A tunnel was then run, but this proved to be too high to drain off the water. Another then had to be commenced, and the work upon the two occupied nearly two years, during all of which time the claim was wholly unproductive. The owners have struggled along with great difficulty, contending with poverty and debt until they could hardly obtain credit for food. Since they 'struck it,' they are independent, and $150,000 is said to have been taken from their tunnel, and rewarded their persevering industry. There are a number of quartz leads in the neighborhood of Placerville which are known to be rich, and will pay well to work.

"The country in the vicinity of Diamond Springs is rather flat or rolling, and the bed-rock, which is a soft, gray slate, lies close to the surface. The mining consists in washing the gravel and a few inches of the rock, in long sluices, worked by companies of from two to four men.

"Eighteen miles southeast of Placerville by the trail, or twenty-five miles by the stage road, is Grizzly Flat. In the lower portions of Cedar Cañon, String Cañon, Lyon's Cañon and Steeley's Fork, a few miners have water sufficient during summer to work their claims, and some of them are doing well. In winter time, when water is plenty, mining is carried on, and good pay is made on the large flat directly back of the town, at Hereford's Ranch, Chris' Ranch, Spring Flat and Meadow's Flat, and in other places in the vicinity. In the flats, the soil and clay above the pay-dirt and bed-rock are about twelve feet in depth, and as there is sufficient fall to the water, it is used in the hydraulic fashion in washing.

"There are several quartz leads on Steeley's Fork of the Cosumnes river, but the gold from them is of poor quality, containing a large percentage of silver and base metal. The amalgam is in summer time worth but six or seven dollars per ounce, and in winter but three; and the gold, when sent to the Mint, produces from thirteen to fourteen dollars per ounce. As the mills are located on the bank of the river which here runs through a steep cañon, the quartz has to be carried down to them from the veins which are near the tops of the hills, a distance of from a quarter of a mile to a mile and a quarter. Some of this work is done by wagons, but two of the mills have cars and wooden railroads, one of which is over a mile in length, and the other three quarters of a mile."

The mining in the immediate vicinity of the town of El Dorado, formerly called Mud Springs, is chiefly in shallow surface diggings. About a mile distant is the Pocahontas quartz lead, of which Mr. Capp says:

"For a couple of years it was worked by Mexicans, whose only aim was to get out the richest of the rock, with as little trouble as possible, and with no regard whatever for the future. In 1854, a Dr. Scott undertook to work the lead, using an apparatus called Bullock's crusher, to pulverize the rock. This is similar in principle and shape to a Chile mill, but much heavier, and was turned by horse-power. Though machines of this kind are said to be used with great success in the eastern States, for want of experience this proved useless here, and after an outlay of $6,000, the experiment was abandoned as a failure, and the machine sold as old iron to a foundry. The vein is about two feet and a half in thickness, crossing a small flat, in nearly an east and west direction, with a dip of about forty-five degrees. It is worked from inclined shafts. The quartz is not much decomposed, but lies in thin layers, and is thus easily broken up and taken out, at the rate of from three-fourths of a ton to double that quantity per day to the hand. The quartz is white, and marked with greenish veins, containing sulphurets of iron, in connection with which the gold is found. The lowest yield, when ordinary rock was ground, has been $18.45 per ton. It is seldom that the gold is visible in this rock, though a number of pieces selected at random in the mine, and from the heaps of rock around the mouth of the shaft, when tested in a mortar, all yielded good prospects.

"Indian Diggings is a mining village twenty-five miles southeastward from Placerville, on the bank of Indian creek. In this district, a belt of limestone, or blue and white marble, rises in ridges through the slate bed-rock, and is in places cut by the water into long and deep channels, some of which serve as natural tail races for the miners, but it oftener renders large amounts of blasting necessary. The claims in the bed of the creek formerly paid well, as the tailings washed down from the hydraulic claims above continually enriched them. In some of the creek claims, in the middle of the channel, deep holes were found, filled with a kind of dirt different from that above it. This was sometimes extremely rich, and in one claim a single panful paid three dollars. Another singular feature connected with these deep places is, that they seem to have subterranean outlets, for in one instance a hundred inches of water poured in for three days, with all the dirt it washed down, failed to have any perceptible effect in filling it up. It was finally stopped with bushes and gravel, and the water turned off. A mile or more above this, in another claim, a similar hole was discovered, and forty inches of water poured in for several hours produced no visible progress towards filling it. Here the miner was in doubt whether there was a rich deposit of gold awaiting him down there, or whether the bottom of his claim had fallen out altogether."

Amador County. § 75. The principal mining towns of Amador county are Butte City, Amador City, Volcano and Jackson, the last being the county seat, and fifty-one miles distant from Sacramento in a south-eastward direction.

"The principal rock around Volcano," says Mr. Capp, "is a hard blue and white limestone, or coarse marble. Under this, and mixed with it, lies the ordinary brown and gray slate, common to all the mining districts. Above the limestone, and forming the tops of the hills, is a light gray, yellow and reddish colored stone, somewhat resembling coarse chalk, but harder. This the miners call 'lava;' geologists, 'tufa.' When some portions of this stone are exposed to the atmosphere, they soon commence to crack, and the corners and angles scale off, and finally are decomposed into a fine light powder, resembling pulverized clay. Other portions of the rock are harder and much more durable, and as they may easily be worked and squared with an ax or sharp stone hammer, they make a handsome and useful building stone. In some of the hills large masses of rock are found, portions of which appear to contain iron, and are full of cells, from one to four inches in diameter, the sides of which are lined with curious coatings of minute crystals. Other portions of this rock are beautifully marked, and resemble flint. All this rock has the appearance of having been melted.

"In the hills behind the town very many claims have been opened, both by tunnels and in the hydraulic fashion, to reach a layer of pay-dirt found under the lava and on top of the limestone and slate. These works have been extremely laborious and expensive, and as there is no regular and well defined lead, while a few of the claims have paid well, others have merely yielded low wages, and the rest have not remunerated the miners for the labor and money expended. The hydraulic claims can only be worked during the spring and winter, when water is plenty, and those miners working tunnels are also unable to wash the pay-dirt they take out in summer time, for the same reason. The amount of blasting required is also large, so that, as a general thing, mining operations at Volcano are carried on under many disadvantages. Many of the hill-sides back of Volcano are composed of rough, loose fragments of limestone, blackened by exposure to the atmosphere and decay. This, together with the conical shape of some of the smaller hills, the curious crystalline and porous structure of a brown and flinty rock which would seem to have been melted, and the presence of large quantities of the rock which the miners call lava, are, I believe, the peculiarities which gave occasion for the name 'Volcano,' which was bestowed on the town. The rock called lava is, however, not at all peculiar to this locality, but is much more plentiful in many other very different looking places. Its appearance and texture also indicate that it is a sedimentary deposit and not a molten rock, and as such it is regarded by all intelligent persons in most other places, though it often goes there by the same name."

Calaveras County. § 76. The principal mining towns of Calaveras county, are Mokelumne Hill, San Andres, Angel's, Murphy's, Vallecito, and West Point. Mokelumne Hill, the county seat, is sixty-nine miles from Stockton.

"Every variety of mining operations has been carried on successfully in the neighborhood of Mokelumne city. The Mokelumne river has been flumed year after year, and thousands of miners have made wages or 'their piles' from its oft-washed bed and banks. Chinamen are the principal miners along this stream during the summer season, and I was informed that they had expended seventy thousand dollars in the purchase of claims, and to secure themselves from molestation in working them. The hill-sides are whitened by the marks of hydraulic mining and the tailings from the numerous tunnels that have been run. There is scarcely one of the numerous ravines and gulches in the vicinity, the bed of which has not been overturned, year after year, since 1849; first, with the butcher-knife and pan or *batea;* then, with the pick and shovel and the rocker; next, with the long-tom; and finally, with the sluices. At last they were abandoned to the

Chinamen with their rockers and the Digger Indian women with their little crow-bars, horn scrapers and tin pans.

"The tall Stockton hill, which lies west of the town, was originally taken up in small claims only a few feet square, and large numbers of perpendicular shafts were sunk in it. Few or none of them paid well, as water interfered sadly with the operations of the miners. Many of these claims were abandoned; and others were finally consolidated into a couple of large tunneling claims. A very lengthy tunnel was then run, at a depth sufficient to insure the requisite drainage, and it has been worked successfully for two or three years back. The old shafts are yet open on the hillside, and the large heaps of white clay and cement by which they are surrounded, remain as monuments of the misdirected industry of the miners of 1852.

"A short distance east of the town commences a high ridge, about six miles in length, which has been taken up, from one end to the other, in tunneling claims.

"The main wealth of the district about West Point consists in its quartz leads, which are so numerous that several of the residents informed me that, starting three miles north of West Point, and proceeding south for a distance of nine miles to the junction of the forks of the Mokelumne, a person would cross a quartz vein in every hundred yards. About one hundred of these veins have been prospected upon the surface, and scarcely any have been found that did not prove to contain gold. As a proof of the richness of the veins of this district, it would be sufficient to state that large numbers of Mexicans and other Spaniards are now working them successfully, although they pay from one dollar to one dollar and fifty cents per cargo of three hundred pounds, to have the rock ground in arastras, to which freight from the leads to the mills along the river has also to be added. Mexicans who do their own work, cannot possibly afford to work rock that does not at least pay three dollars per cargo, or twenty dollars per ton, and in fact they seldom do work rock that pays less than six dollars per cargo and forty dollars per ton.

"There is very little slate in this district, and nearly all the quartz veins are encased in granite, which is usually much decomposed. Occasionally, the granite appears to 'pinch' the quartz leads until they become very thin; but by tracing them on further, or downwards, they again swell out to their original size, and sometimes bulge out beyond it. In such places, and at the intersection of small veins, very rich deposits of gold are frequently found, which, from their narrowness and the depth to which they extend, the Spaniards call *clavos* or nails. In other places, the granite becomes somewhat mixed with the quartz, which is what the Spaniards call *bora*, or in speaking of the quartz, they say it is *emborascado*."

Angel's is one of the first quartz mining places in the State. The principal portion of the rock "is of a greenish and gray color, and contains large quantities of the sulphuret of iron. Mixed with this, are streaks and veins of white quartz or limestone. The sulphurets are found, either in irregular, bright, crystalline masses, or small threads and veins. Some of these veins are as much as eight inches in thickness. In other portions of the green rock, the sulphurets are scattered all through it, as separate and minute square crystals. The whole formation will probably become one solid vein when any considerable depth is reached; but near the surface it is cut up into separate veins by streaks and wedges of slate, which do not appear to contain any gold. These streaks of slate are from a few inches to several feet in thickness. The poorer portions of the rock contain from twelve to sixteen per cent. of the sulphurets, while the richest are nearly pure crystals, among which the gold is seen shining in small particles and scales."

The mines about Murphy's are chiefly of the placer kind, with much deep digging.

At Copperopolis, in the south-eastern part of this county, are rich veins of copper ore, which have lately been opened, and in which men are now regularly employed at mining.

Tuolumne County. § 77. Tuolumne is the most important mining county in the Southern mines. The principal mining towns are Columbia, Sonora, Shaw's Flat, Chinese Camp and Big Oak Flat. Sonora, the county seat, is sixty-five miles from Stockton. There has, probably, been no placer district in the State so rich, within the same extent, as that within five miles of Sonora. No other part of the State has furnished so many large nuggets. Two very large ditches furnish water to the miners near Columbia and Sonora; but of late, much damage has been done to one of these ditches, by malicious persons, who have injured the works so that much of the supply of water has been cut off. One of the greatest wonders of the Californian mines is Table Mountain, in this county. It is a long hill or mountain of basalt, from one hundred to seven hundred feet high, from a hundred to five hundred yards wide, and thirty miles long. Its course is serpentine; its general direction, south-west; its eastern end is in Calaveras county, its western near Dent's Ferry, Amador county, while the great body of it is in Tuolumne county. It passes near Sonora. Its surface is nearly level, gradually descending towards the westward. General opinion has decided that this mountain was formed by a stream of lava, which ran down the bed of an ancient river, filling the river bed to a level with the banks; and that, in the course of time, the banks were carried away, leaving the hardened lava rising above the adjacent country like a mountain. The basalt is certainly of volcanic origin, and no other explanation is given of its narrow course and steep banks. Besides, the existence of an ancient river bed under the mountain cannot be denied. There are the beds of gravel and clay; the flat stones all pointing down stream; the water-worn bed; the remains of trees and fresh water mollusca; the gold which was collected by the water; and the tributary streams. This old river bed is extremely rich. In one place, a tract one hundred feet square, yielded $75,000 in gold, or $7.50 to the square foot. The pay-dirt is from six inches to six feet deep. The richness of the dirt was discovered by following a lead which had been one of the small tributaries of this ancient river. This lead did not take the miners fully into the great wealth of the old river bed; but it suggested the existence of a vast treasure there, and they cut a tunnel through into the channel; for the only way of reaching the gold, except at one or two places, where little streams emptied into the large river, was to cut tunnels through the rim-rock, which rose along the edge of the ancient river, above the channel and above the level of the ground along the side of the mountain. The facts of the existence of the ancient river bed, and its richness, were discovered and made known in October, 1854, and created a great excitement. Some persons made fortunes, and others lost them. In one case, $100,000 were spent in cutting a tunnel or drift, and then it was useless because too high, and the company had to cut another one. The tunnels to reach the pay-dirt under Table Mountain, are from 600 to 1200 feet long. Of late, the owners of claims in Table Mountain are getting the dirt out through inclined shafts, which start into the hill above the top of the rim-rock, so that they are spared the expense of blasting. The water is drained out by the older tunnels.

Eastward from Sonora are some of the richest quartz veins in the State, and one claim, called the Soulsby lead, is especially noted for the very rich quartz which it has furnished in times past. Very little has been said of late about this claim; but in 1858 it attracted great attention. The Sonora *Democrat* spoke thus of it, when the mill attached to the claim commenced operations:

"Last Saturday the amount of retorted gold brought to Sonora, from the week's crushing by this mill, was sixty-five pounds. After the above result was known, a remaining balance of amalgam was retorted, yielding over nine and one-half pounds of gold. For the week's operation, therefore, the sum of seventy-four and one-half pounds of gold, worth $15,000, has been brought into town from the claim of Street & Soulsby.

"The number of hands employed in this claim, in raising quartz, wheeling and grinding the same, etc., is nine. The cost, as yet, of raising and delivering the rock at the mill, is less than four dollars per ton. The mill has been in operation just nineteen days, and the aggregate of gold already saved from this operation, during said time, amounts to two hundred and fifteen pounds and three ounces, worth $40,000.

"The claim of Street & Soulsby is 2,400 feet on the line of their lode. They have explored this lode at various points, and to various depths, and in no single instance has rock been raised showing any diminution of richness. Now, the vein at every point, exceeds the thickness of two feet. It becomes wider or thicker than this as their explorations descend. Also, the rock yields, under the imperfect process of their stampers, more than $200 per ton. Now, take the above data, which are below the truth, in the richness and width of the vein, and, by the assumption that the same quality of rock occurs at the depth of one hundred feet, we arrive at the following result: Length of lode, 1,200 feet; depth, 100 feet; thickness, two feet; 1200x100x2—240,000 cubic feet of gold-bearing quartz. Supposing the specific gravity of this quartz to be two and one-half, (we think it to be nearly three) we find that twelve cubic feet thereof will weigh one ton. Therefore the above estimate, by a little simple arithmetic, shows the quantity of quartz, within one hundred feet of the surface of the said lode, to be twenty thousand tons. As this rock yields $200 per ton, ($300 is nearer the truth) it presents, on the above estimate, the astounding value of $4,000,000."

Large districts in Tuolumne county, including the mining grounds at Chinese Camp, rich as surface diggings, had a shallow bed of red gravel, from six inches to six feet deep, lying upon brown, yellow and greenish slates, the strata of which were perpendicular. Here and there, pieces of the slate would project above the gravel, looking like grave-stones. It has been suggested that the richness of this ground might be explained, by supposing that a great body of auriferous matter had been swept, by water, over these slates, which caught the gold, as a riffle does, by its rough surface.

Mariposa County. § 78. Mariposa county was once very rich in shallow placer diggings; but now its chief wealth lies in the quartz mines in Bear Valley, on Fremont's ranch. The Benton quartz mill is the largest in the State, and has forty-eight stamps, and has a capacity to crush seventy-two tons of quartz per day. There are four quartz mills on the ranch, with ninety-one stamps in all. A railroad, four miles long, conveys the rock from the vein to the mills. The various mills on this ranch now yield $60,000 per month.

"There are," says a correspondent of the *Alta California,* "about twenty known quartz leads on the grant, nearly all of which communicate; but the 'Josephine' and 'Pine Tree' are the principal—in fact, the latter is termed the mother of the quartz leads. These veins run about north-west, and the Pine Tree is traced from near Mount Ophir, all through Bear Valley, and thence along the ridge of Mount Bullion, to the Merced river. Here and there the quartz seems cut off, or spread into several smaller veins, by the interference of different rocks; and, as a general thing, these smaller veins possess some richness. After several disruptions, these smaller parallel veins appear to unite into a compact lead, and even as far as Peña Blanca, twenty miles from the Merced, their out-croppings form the marked crest of that locality. Although the Josephine vein is, apparently, a split from the Pine Tree or mother vein, it apparently belongs to a later geological period, as is evidenced by the character of their relative ores. Both rocks are of flinty hardness, and white, with here and there streaks of dark gray, nearly black, from the deposit of sulphurets.

The Pine Tree quartz presents its metals mostly in an oxydised state, the iron looking rusty or red, the copper as red oxide, or blue and green carbonate, which gives a variegated and fantastical appearance to the rock, much of which is truly beautiful. Instances of perfect crystallization have been known but rarely, although now and then imperfect formations, through which are traced fine lines of gold and copper, have been met with. The Josephine rock, unlike the Pine Tree, presents all the sulphurets in their original metallic luster, and is free from any marks of atmospheric action; in fact, this and the Princeton vein contain the most of their metal in the form of sulphurets.

" The peculiarity of the Josephine vein is the streaks, layers or strata of sulphurated rock found therein. These run, in various thicknesses, as far as the work exhibits, and four different ones have been found. The most westerly of these layers, which is struck about one hundred and eighty feet from the mouth of Black Drift, is from twenty to twenty-two feet in thickness, the rock paying from forty-five to fifty dollars per ton. The sulphurets are somewhat coarse, bright in color, appearing mostly in scales, with no free gold apparent amongst them. The second streak, and the richest ore, is struck about two hundred and seventy-five feet from the mouth of the same drift, is about fifteen feet in width, composed of compact and fine sulphurets, carrying with them a large amount of free gold, which is plainly visible to the unaided eye. In the drift, by the light of a candle, the rock glistens, as if faced with diamonds. This rock, by careful treatment, pays from five hundred to two thousand dollars per ton. The rock or quartz is pearly white in color, with streaks of dark bluish gray, very hard, softening on exposure, and partaking more of feldspar than the rock which is taken from the more southerly leads, in which lime appears. Seventy feet beyond the second streak, the largest strata is found. This is from thirty to thirty-five feet in width, and of the same character as the first, differing, however, somewhat, as the first ten feet on the westerly side contain a small quantity of free gold. Twenty feet further on, another streak has been struck, the thickness of which has not yet been ascertained. The sulphurets here are in cubes, containing but little gold. All these streaks run through the vein perpendicularly, at an angle of eighteen degrees, tending alike towards the junction of the Pine Tree and Josephine veins, where the quartz is at least some thirty feet thick, and where a vast deposit of very rich ore will undoubtedly be found.

"The Princeton, or Ridgway & Steptoe vein, is located seven miles from Bear Valley, and two miles from Mount Ophir, being two thousand seven hundred and eighty-two feet above the level of the Merced river. Two drifts are being run, which are connected with the surface by four shafts. The vein averages from three and one-half to four feet, and is cased in friable slate, requiring great care in working, and much expense in timbering. At a depth of from fifteen to twenty-five feet, the rock pays from fifteen to twenty dollars; at the depth of from sixty to eighty feet, it pays from thirty to forty-five dollars. The main shaft is opened one thousand feet, from the western to the eastern shaft. In the eastern shaft, at a depth of one hundred and sixty feet, the rock pays forty-two dollars; in the Green shaft, at a distance of six hundred and twenty-five feet, at a depth of one hundred and twenty-six feet, it averages thirty dollars. At the Ridgway shaft, two hundred feet further distant, at eighty feet depth, it pays thirty-five dollars; and at the New Whim shaft, one hundred and seventy-nine feet still further, at fifty feet depth, the rock pays thirty dollars. A portion of the rock is filled with sulphurets, richly sprinkled with free gold. The peculiar feature of the rock is, the beautiful laminated appearance of the gold, which, polished as bright as if by the hands of a goldsmith, is found attached, in single or double leaves, to the white quartz, by a single thread."

A large portion of the richest mining land in the county is owned by J. C. Fremont. Mariposa town, the county seat of Mariposa county, is ninety-five miles from Stockton, in a southeastern direction.

There are a few miners in Fresno, the most southern county of the "Southern mines," but there is nothing peculiar about the diggings there.

We have thus given a general idea of the main features of the main or Sacramento gold mining district of California, and shall devote less space to the other districts, which are of less importance, and possess the same general character.

Shasta District. § 79. The next in importance of the gold mining districts of California, is that at the northern end of the State, including the counties of Shasta, Siskiyou, Trinity, Klamath and Del Norte. This may be called the Shasta district. It is nearly one hundred miles square, and is watered by the head waters of the Sacramento river with its tributaries, Clear creek, Cottonwood creek, Churn creek, McCloud river and Pitt river, and by the Klamath river with its tributaries—Trinity river, Scott river, Shasta creek and so on. Most of the miners in this district are at work in placer diggings of various kinds; quartz mills are not so numerous or so large as in the Sacramento district.

Shasta City, the county seat of Shasta county, is 176 miles from Sacramento; Yreka, the county seat of Siskiyou, is 248 miles from Sacramento; Weaverville, the county seat of Trinity, is 210 miles from Sacramento; Orleans Bar, the county seat of Klamath, is 250 miles from Sacramento; and Crescent City, the county seat of Del Norte, is 300 miles, by sea—the route of all the trade—from San Francisco. In Klamath county, on the sea shore, is the famous Gold Bluff, that caused the great excitement in 1850. There is here a high bluff bank, at the foot of which is a sand beach, much of which lies between high and low tide marks. This beach has been made by the washing away of the auriferous bluff, and the particles of gold are left in the sand; but they are so very fine, that they move about with the sand, and are scarcely to be caught by the miner. A mining shaft six hundred and fifty feet deep, the deepest in the State, was dug about a mile north of Weaverville. It was undertaken to discover whether gold could not be found on the bed-rock, and most of the labor was paid for by public subscription. Neither bed-rock nor gold was found. The next deepest shaft in the State was sunk at Sutter creek, Amador county, three hundred and fifteen feet deep, in a paying quartz vein.

The placers of the lower portions of the Klamath and Trinity rivers contain much platinum, iridium and osmium, amounting in some spots to one half as much as the gold. The three metals are usually found together, platinum forming forty per cent. of the three. All are whitish, silver-like metals, of nearly the same specific gravity with gold, from which, therefore, they cannot be separated by washing. It is so difficult to separate them by processes used in laboratories, that the Mints will not receive dust containing either of these metals, though occasionally a bit will slip into the coin, where it appears as a whitish or yellowish speck in the gold; for these white metals do not melt and mix freely with the gold. The only convenient mode of separation, is with quicksilver which does not unite with them, all the gold being caught in the amalgam, which gathers in large lumps, and may be readily separated from the independent particles of other metal. The iridium and osmium are often found mixed together, and are then called irid-osmium. There is no demand, and no regular price for platinum, iridium or osmium in California.

Kern River District. § 80. The Kern river district lies about latitude 35° 30', in the south-eastern corner of the Sacramento basin. The mining done there now is chiefly in quartz, and the arastra is the favorite instrument for pulverizing and amalgamating. White river is about forty miles north of Kern river, and belongs to the same district. Along this river, says a newspaper correspondent, are "immense boulders of granite, that seem to have been by some convulsion toppled from the hills, and to have rolled down the pitches to the commencement of the level ground, where they are piled upon each other, to a great height. The boulders vary in size, being from ten to fifty feet in diameter. Being round, there are, of course, large crevices between them, in different directions, that the miners call caverns, and the gold-hunters travel through these passages or caverns for a great distance, seeking for an opportunity to get at such portions of the bed-rock as there is ample space to work. When they find such a spot, they are compelled to let themselves down from one boulder to another, till they reach it. Traveling through these passages is quite a labyrinth to the uninitiated. On these placers not covered, the water in the raining season has, by constant action for years, and by the boulders acting as dams, eddied around and worn into the bed-rock, wells from three to ten feet in depth, and from four to six feet in width. The dirt found in these is very rich, all the way down, though such is not the case with holes of the kind in rivers. Much of this gold is mixed with crushed quartz."

About latitude 34° 30', in the coast mountains, are the San Francisquito and San Gabriel diggings, though separated by half a degree or more of longitude from each other, and each about forty-five miles distant from Los Angeles, which is four hundred and eighty miles from San Francisco.

In latitude 38° east of the Sierra Nevada, are the Mono

and Walker mines, little districts which have a couple hundred miners, in the summer but none in the winter.

Fraser District. § 84. The mines of Fraser river extend from latitude 50° to 58°, and they are between one hundred and two hundred miles distant from the coast. Some auriferous bars are found along the river, below the point where it breaks through the Cascade range; but the great body of the diggings is east of those mountains. The mining is all done in shallow places; no rich quartz leads have been discovered, or at least, none are extensively worked. The river is navigable for steamboats, from the ocean to Fort Yale, a distance of one hundred and ten miles. From there, for a distance of fifty miles up stream, the river is broken by numerous falls and rapids, which may be passed over in canoes, at some stages of water; but always with some difficulty and danger. The first tributary stream that bears gold in its sands, is Thompson's river, which pours down a large body of water from the east, and empties into the Fraser, two hundred miles from the sea. Above Thompson's river, on the west, are Bridge and Chillicoaten rivers; on the east, the Quesnelle and Cariboo rivers. All these are gold-bearing. Sixty miles above the mouth of Thompson's river, is called the "Canoe Country;" beyond that, to the north, lies the "Balloon Country;" and beyond that, the "Cariboo Country;" terms objectionable, because of their indefinite character; but still generally used. Access to the mines, up the Fraser, is very difficult and costly; and there is another route, by the Lillooet (or L'Alouette) river. The following are the distances by this route:

From Fort Langley to mouth of Harrison river	35 miles.
Across Harrison lake	40 "
Up Lillooett river	50 "
Across Lillooet lake	18 "
Small river	3 "
"Long Portage"	30 "
Across Third Lake	15 "
Portage	5 "
Across Lake Anderson	12 "
Last portage over to Fraser river	8 "
Total	216 miles.

Steamers run regularly from Victoria to Fort Hope, and from San Francisco to Victoria.

The following figures show the distances from Victoria to the Big Falls of Fraser river:

From Victoria to Fort Langley.. 80 miles.
" Fort Langley to Fort Hope.. 60 "
" Fort Hope to Fort Yale... 15 "
" Fort Yale to Thompson river......................................110 "
" Thompson river to Big Falls, Fraser river.................... 75 "

    Total..340 miles.

The following are the distances from Portland to the Fraser, by way of the Dalles:

From Portland to the Dalles by steamboat............................. 90 miles.
Thence by wagons or pack animals to Chickitat valley............. 25 "
" to Simcoe.. 40 "
" " Yakima... 30 "
" " Columbia above Priest Rapids............................ 35 "
" " Wenatchee... 65 "
" " Crossing Lake Chelan.. 35 "
" " Methow.. 30 "
" " Okinagan... 20 "
" " Mouth of Similkameen (here leaves Rock creek route to
    the north-east).. 60 "
" up Similkameen... 60 "
" to Nicholas valley.. 30 "
" " Fort Thompson... 45 "
" " Mouth of Kamloops... 20 "
" " Bonaparte valley (here route up Fraser river intersects).. 35 "
" up Bonaparte river... 30 "
" to Goose lake.. 25 "
" " Bridge river... 25 "
" " Nine Mile lake.. 28 "
" " Williams lake... 30 "
" " Mud lake (here fork leads off to Fort Alexander)........ 30 "
" " Beaver lake... 25 "
" " Forks Canal river... 20 "

Total distance from Portland..823 "
From the Dalles..733 "
From Dalles to the intersection of Fraser river route............520 "

Rogue River. § 82. Part of Rogue river valley, in southern Oregon, is rich in gold. The country is rugged, and much of it covered with dense forests, which interfere greatly with the labors of the miners, but protect the mineral deposits from speedy exhaustion, and secure some of their profits for future generations. The surface diggings of southern Oregon will, for this reason, probably last longer than most of those in California. Some very rich quartz lodes have been found in Rogue river valley. The two most noted ones are the "Ish" lead and the Applegate lead. The Ish lead, at Gold Hill, produced $20,000 in one week in 1859, and in a year paid about $200,000. Work in the Applegate lead, near Applegate creek, was commenced in June, 1860, with an arastra, and for six months the lode yielded only forty dollars per ton, but in January, 1861, the workmen came to very rich rock. In February of this year, the arastra was producing about $5,000 per week clear profit. Ten tons of picked rock contained $23,-520 of gold. Platinum, iridium and osmium are found in the western part of Rogue river valley, to almost as great an extent as in the Klamath placers.

Upper Columbia District. § 83. The valley of the Upper Columbia contains a large extent of auriferous ground, rich enough in many places to pay for. working. The lowest point where gold has been discovered is near Fort Walla-Walla, about 350 miles from the ocean, and from that point upward for 500 miles or more, along the course of the river, gold has been found in its banks and bars, but not enough to pay for working, except in a few places. The Yakima, the Pisquouse, the Wenatchee, and several other streams running down the eastern slope of the Cascade mountains to the Columbia, have had some rich bars. Gold is also found on the Okanagan river, and along Okanagan lake. The mines of Rock creek and Kettle river, in the valley of the Upper Columbia, are highly spoken of. Rock creek is about 300 miles distant from the Dalles, in a northward direction. It is a tributary of Kettle river, which empties into the Columbia near Fort Colville. Boundary creek is another tributary of Kettle river, and has some rich bars, where miners are now at work. All the country in this vicinity is auriferous. The amount made per hand per day varies, of course, as in all mines; but the wages for good laborers is four dollars per day. There have been cases where men have made more than a hundred dollars a day with a rocker. They were mining on Boundary creek, part of which is in British Columbia. A nugget worth $560 was found in September, 1860, on Rock creek, which is entirely in British Columbia. All the other auriferous branches of Kettle river are in Washington territory. Kettle river itself is about sixty yards wide. Much mining will be done on this stream and its tributaries. There is good grass and rich soil in the basin of Kettle river. Two hundred persons are now in the Rock creek mines. The Similkameen, which lies in British Columbia, but is a tributary of the Columbia river, has some of the richest mining ground on the coast. Near Fort Colville, distant by the course of the river about 800 miles from the ocean, miners have been at work since 1855. The "Colville mines" are on the Pend-Oreille (pronounced "pon-du-ráy") river, a large stream, which, after flowing in a west-north-west course, empties into the Columbia about ten miles south of latitude 49°. Its body of water may be approximately expressed by the formula of eighty yards wide, four feet deep, and a current of six miles an hour. It gets very low, however, in the winter. The Colville mines commence on the Columbia river, near the mouth of the Pend-Oreille, and extend up the latter stream twenty miles, to a tributary called Salmon river, and up Salmon river four or five miles. The mining ground on the Pend-Oreille is all in a deep cañon, except at its mouth, where there is a flat. The diggings are on bars, and pay from three to six and eight dollars per day. The best diggings have been found on a flat half a mile long and several hundred yards wide, at the mouth of the Pend-Oreille, which flat was once undoubtedly a bed of the river. The bed-rock has not been reached at any place in these diggings. The mines were discovered in 1855, by Jo Morell, a Canadian. The bars of the Columbia are auriferous for seventy-five miles down from the mouth of the Pend-Oreille to the Spokan river, which flows from the east.

The Columbia river is navigated by steamboats from its mouth to Walla Walla, a distance of 360 miles. The river is navigable above Walla Walla, but there are no vessels running regularly. Parties going to the mines in the basin of the Upper Columbia sometimes begin their land journey at Walla Walla, but more frequently at the Dalles, which is two hundred miles from the ocean. From the Dalles to the Wenatchee is 150 miles; to the Similkameen, 250; to Fort Colville, 350. The traveler from the Dalles to Rock creek goes sixty-five miles to the Simcoe Indian Reservation, over a fine wagon road; thence to Atahnam river, twenty miles; to Mission creek, eight miles; to the Nahchess river, seven miles; to the Wenass river, five miles; to the Yakima river, eighteen miles; to the Columbia, twenty-five miles; up the Columbia to the Wenatchee, ten miles; to Rock creek, seventeen miles; to Chelan lake, eighteen miles; to the Methow, sixteen miles; to Okanagan lake, eighteen miles; to the mouth of the Similkameen, forty-five miles; to Rock creek, twenty miles. The road is good for pack mules the whole distance, and grass is abundant.

It is reported that rich gold diggings have been found in the valley of the Clear Water river, on the western slope of the Bitter Root mountains, and in the basin of the Columbia river, about one hundred and fifty miles south-eastward from Fort

Colville, and the same distance eastward from Walla Walla. The *Portland Times* says of these mines:

"They are well elevated in the mountains, which are densely covered with pine timber, except in small patches where cammas and grass prairies and swamps are found. The climate is colder than in the valleys below; but, nevertheless, so mild as not to prevent work this winter.

"The mines are on the Nez Perces Reservation, as claimed by the authorities. The Nez Perces Indians are numerous, wealthy, and skilled in war, and some of them instructed in the arts of peace. They have always been friendly to the whites, and are free from the little vices which characterize many other tribes. They number about 4,000 souls, about one-fifth of whom are warriors. Their wealth consists chiefly in horses and cattle. The Government has an agency among them, and is now instructing many of them in the art of agriculture. The agency or head-quarters of the Reserve is on the Clear Water, near the junction of the Lapwa, and about one hundred miles from Walla Walla, by the road usually traveled.

"The route from Portland to these mines, is by steamboat to the Cascades; thence (after crossing the portage) by steamboat to the Dalles; thence by stage to the Des Chutes; thence again by steamboat to Old Fort Walla Walla; thence by stage to the New Fort and the town; thence by wagons or pack animals to the Red Wolf crossing of Snake river; thence mostly by pack animals up the Snake river, and Clear Water and its northern branch to the mines.

"A good wagon road can be built, at but little expense, to within fifteen or twenty miles of the place where the miners are now at work. The grass and water are good and abundant to within about forty miles of the mines; thence the timber is dense and but little grass, save in small openings on the creek bottoms. No part of the route is obstructed with rock, so as to interfere with pack trains. It is hilly from the crossing of the Snake river to the forks of the Clear Water, and through an open pine timbered country; thence to the mines, the ascent up the fork of the Clear Water is more rugged and the timber more dense.

"The following is the table of distances:

From Portland to the Cascades	50 miles.
Cascade Portage	2 "
Thence to Dalles	38 "
" " Des Chutes	15 "
" " Old Fort Walla Walla	130 "
" " New Walla Walla Town and Fort	30 "
" " Toucha	26 "
" " Tuchanon	20 "
" " Red Wolf Crossing of Snake River	30 "
" " Mouth of Clear Water	10 "
" " Indian Agency	15 "
" " Forks of Clear Water	30 "
" " the Mines	40 "
Total from Portland, Oregon	436 miles."

Washoe. § 84. In latitude 39° 30′ north, and longitude 119° 45′ west of Greenwich, in the basin of Carson river, and very near the eastern base of the Sierra Nevada, lies the Washoe silver mining country, about one hundred and seventy-five miles distant, in an east-north-eastward direction, from San Francisco. The name "Washoe," which in the beginning of 1859 was confined to a small valley west of the argentiferous region, is now given to all the silver bearing country in its vicinity; a district that may be twenty miles square, although all the rich lodes as yet opened are within a circle whose radius is not more than four miles.

Washoe is part of Nevada territory, and of the great interior basin which sends none of its waters to the sea, but swallows up all its rivers and brooks in its own sands. Among its rivers, in the metalliferous regions, are the Carson, Walker and Mono, each of which has its lake or sink. The land has an elevation of 4,500 feet above the sea in its lowest parts. Its surface is broken and mountainous; its soil dry, sandy and sterile; its vegetation scanty, scrubby and desert-like, and its climate fickle in summer and severely cold in winter. Washoe is frequently spoken of as being on the "eastern slope of the Sierra Nevada," but the term is incorrect; the slope eastward being very short as compared with that on the western side of these mountains.

Washoe is divided into about a dozen mining districts, of which the principal are the Virginia, Gold Hill, Devil's Gate, Flowery, Argentine and Silver Star.

The Virginia district includes Virginia City and the Comstock Lode, on which are situated all the very valuable silver claims in Washoe. Virginia City, the most important mining town in Nevada, has now a population of about three thousand. It is situated in a little basin, nearly circular in shape and about a mile across. There are about twenty brick and stone houses, the remainder are of wood. Immediately west of the town and almost within its limits, is that part of the Comstock Lode owned by the Ophir Company. Running southward from Virginia City, is Gold Cañon, five or six miles long, the sides of which are rich in mineral wealth. The Comstock Lode is found about two hundred feet above the level of the basin of Virginia City, on the eastern slope of a very steep, bare, rugged, rocky hill, seven hundred feet high, forming one side of Gold Cañon. The Comstock Lode is from fifteen to thirty feet wide, and its ore is very rich for a distance of a mile. The lode, in its rich portions, is owned by the Ophir, Mexican, California, Central, Gould and Curry, Chollar, and other companies. The rich part of the Comstock lode is about a mile and a half long. The vein stone is a white quartz, between walls of porphyritic greenstone and amygdaloid trap. The vein has a direction nearly north and south, and a slight dip to the west. All the companies above named as owning rich claims on the Comstock lode, have sunk shafts and taken out rock in considerable quantities, and most of them have erected or are erecting mills.

A correspondent of the San Francisco *Bulletin* wrote in the midst of December, 1860, thus of the Mexican claim :

" Work was commenced on this mine by a sinking a shaft, in size about fourteen by eight feet. This shaft was sunk on the vein of ore, which was struck some ten or fifteen feet from the surface. The inclination of the vein is sufficient to allow rude steps to be cut on the lower side of the shaft. The ore and quartz is taken from the mine up these steps on the backs of Mexicans. The ore is carried in a sort of basket made of raw hide, in shape like a peach basket, but about a foot higher. A strap attached to the top of the basket passes around the laborer's forehead, allowing the loaded vessel to lie on his back. With this arrangement, experienced Mexicans take enormous loads up places that an inexperienced person would find difficult to ascend or descend without any load.

" When this shaft was down some forty or fifty feet, drifts were run north and south of the vein, the full extent of the claim. From these drifts other shafts are sunk at suitable distances apart. Again, at proper depths, other drifts are run, etc. In this manner, at equal distances, pillars of ore are left standing to support the mine, thereby saving, in a great measure, the necessity of timbering. These pillars are not like timbers, liable to rot, but render the mine permanently secure for any length of time. If at any future period there is no further use for them, they can commence at the bottom to take them out, and let the mine fall in.

"During the past season this company have been erecting a stamping mill, with suitable works for amalgamating silver. The building is ninety feet long, and forty feet wide. It is situated a few hundred yards east of the mine, in the outskirts of Virginia City. For crushing, there are four batteries—in all, sixteen stamps, set in a line, and lifted by cams on one shaft. The ore is crushed dry. It falls from the batteries through fine wire-cloth to conveyors; they take it to elevators, which deposit it in the upper story of the building. For amalgamating, twenty German barrels are arranged. These barrels are capable of working ten tons in twenty-four hours."

The Virginia district is strictly a silver mining district, for although gold is found there, yet it is found only among silver ore, the great value of which is in the silver.

Southward lies the Gold Hill district, the chief town of which, named Gold Hill, is only one mile distant from Virginia City. The wealth of this district is chiefly in gold quartz, found along the sides of Gold Cañon, which runs southward from the basin of Virginia City. The bottom of this cañon was once very rich in placer gold, and the auriferous quartz here is very abundant and rich. Indeed it has been asserted that the vicinity of this place is richer in gold-bearing quartz than any district of equal extent in California. The town of Gold Hill is in the cañon, and has a population of about five hundred.

Three miles south of Gold Hill, and also in the cañon, is Silver City, which has a population of five hundred. This place is in the Devil's Gate district. The claims are mostly gold claims. The three towns are only four miles apart, and will in time become almost one continuous town, strung along a narrow, rugged, steep cañon.

There are three stamping mills at Virginia City; one owned by the Mexican Company, another by the Central, and a third by the Gould and Curry. The Ophir Company have no mill at their mine; but they are erecting very extensive works in

Washoe Valley, twelve miles distant, to which place they have cut a fine road. It is said that the ore already taken out of the Ophir claim and ready to be put through the mill, is worth $1,500,000.

At Gold Hill there are two mills; one owned by Harris and Conover; the other, owned by a company, is called "Paul's Big Mill." It is named after Almarin B. Paul, Esq., who erected it and is its superintendent. The *Territorial Enterprise* newspaper, published in Carson City, thus described the mill, in November, 1860:

"The ground on which the mill is situated is five hundred feet front by three hundred in width, all used for mill, furnaces, buildings, stables, etc. The size of the mill is, one hundred and ten feet long, by seventy-five feet in width; eighteen feet high, to the eaves; and contains one hundred thousand feet of lumber. Every department in the building is independent of each other. To better judge of its size, we will give each department separate: Battery building, twenty-five by seventy-five feet; quartz room, twenty by seventy-five; amalgamating room, forty by fifty; clearing-up room, fifteen by twenty; melting room, fifteen by fifteen; assaying room, fifteen by fifteen. The mill contains eight of Howland's rotary batteries, eight stamps each, in all sixty-four. When all in, it will contain forty-eight amalgamating pans, Knox's patent. The engine is of sixty horse power.

"The material is taken from the batteries for amalgamation, by cars and railways. The metal will be turned out only in bars. The necessary machinery requisite for cleansing, separating and retorting the metals, contain one hundred and twenty tons of iron, brought over the Sierra Nevada at an expense of eight, ten, twelve and sixteen cents per pound.

"The expense of such mammoth operations here, may be inferred from the fact that the company have expended between $25,000 and $30,000 for freight alone. Employment has been given to about fifty men per day. At one time, one hundred men were engaged, day and night, on the machinery. The expenditure of this company, with everything complete as designed, furnaces, etc., will be about $200,000."

At Silver City there are nine mills, and on the Carson river, also within the limits of the Devil's Gate district, there are four mills.

The Flowery is east of the Virginia district, and contains

THE MINING DISTRICTS. 109

some rich gold leads, among which the Lady Bryan and Rogers are prominent.

South of the Flowery is the Silver Star district; and west of the Virginia is the Argentine district, in neither of which have any very rich lodes been discovered as yet.

Virginia City, Gold Hill and Silver City are the only mining towns of note in Washoe. Carson City, sixteen miles southward from Virginia City, is a place of fifteen hundred inhabitants, and is the principal trading town out of the immediate vicinity of the mines.

The following figures give the distances from Sacramento to Virginia City:

Sacramento to Placerville	45 miles.
Placerville to Junction	15½ "
Junction to Brockliss Bridge	2½ "
Brockliss Bridge to Strawberry Valley	26 "
Strawberry Valley to Slippery Ford	1 "
Slippery Ford to Johnson's Pass	7 "
Johnson's Pass to Lake Valley	2 "
Lake Valley to Luther's Pass	4 "
Luther's Pass to Hope Valley	2 "
Hope Valley to Woodford's	5 "
Woodford's to Genoa	20 "
Genoa to Carson City	14 "
Carson City to Virginia City	18 "
Total	162 miles.

Esmeralda. § 85. Esmeralda lies about one hundred miles south-south-eastward from Carson City, and fifteen miles northward from the eastern point of Mono Lake, in the basin of Walker river. There is a good natural road for wagons, through valleys and over low passes, from Carson Valley to the entrance of the Esmeralda mountains, and thence the road is rough for a distance of ten miles. There is a toll road from Walker river to Esmeralda, and a free road from Monoville to Esmeralda. The Esmeralda mountains consist of broken ridges of eruptive rocks, chiefly trap and basalt, covered in many places with volcanic scoriæ. The general course of these ridges is north and south. Bunch grass is abundant; and there are extensive forests of scrub pine, pitch pine and nut pine, (*piñon*) all good for fire-wood, but not valuable for building. Very good water, sufficient for domestic purposes and for the use of steam mills, is found near the mines.

The rich district is supposed to be within a circle five miles in diameter. The elevation of the place is about five thousand feet above the sea.

The main lode, called the Esmeralda, runs north and south, with a width of from thirty to sixty feet of quartz running through porphyritic greenstone. The lode dips slightly to the west, and in some places, projects fifteen or twenty feet, like a high wall, above the adjacent land. The ore is a vitreous sulphuret of silver, with very little gold. Most of the other leads in the vicinity run east and west, or, at least, vary considerably from the north and south direction of the Esmeralda lode. These other leads are many of them rich in gold.

The town of Esmeralda lies at the base of the hill, on the side of which the claim of the same name lies. Two miles further north, upon a better place for a town site, is the town of Aurora, which now boasts one hundred cabins and tents.

The trade of Esmeralda will probably be done with Placerville, to which place the distance is about one hundred and fifty miles. Communication may be had with Stockton, by the Big Tree route, a distance of one hundred and forty miles; and with Coulterville, by the way of Mono and Yosemite, a distance of one hundred and twenty miles.

Coso, etc. § 86. The Coso silver mines lie about one hundred and twenty miles eastward from Visalia, in the great basin of Utah. It is a barren district, where wood and water are scarce. Rich specimens of sulphuret of silver, and argentiferous copper ore, have been shown in San Francisco, said to have come from that place. There are few miners residing there as yet, and our information about the district is not exact or full.

It is reported that rich leads of silver ore have been found near the Mountain Meadows, in the western part of New Mexico, and also at Silver mountain, in the same Territory; but these are rumors, in which little trust can be placed. The Silver Mountain is said to be twenty miles south-westward of Las Vegas, on the road from San Bernardino to Salt Lake, and its ore to be rich argentiferous galena.

Arizona. § 87. Arizona is a district about fifty miles wide, from north to south, and six hundred long, from east to west, lying south of the river Gila, between the thirty-first and thirty-third degrees of north latitude. It is a sterile country, made up, chiefly, of barren sands and bare rocks. It is rich in gold, silver and copper. Its gold is found in places which cannot be worked for want of water. Its silver mines were discovered about the beginning of the last century, at which time there was a considerable Spanish population in the country. It is said that at one time a hundred silver mines were worked; but about the middle of the last century, the hostilities of the Indians disturbed the labor of the miners, and in 1820 nearly all the Spaniards had left the country, and the business of mining had come to an end.

The most important mine of Arizona, is that of the Sonora Exploring and Mining Company. The mine, called, usually, "The Heintzelman Mine," from the President of the company, lies in the Cerro Colorado, thirty miles from Tubac. The ore is an argentiferous galena, which, in picked specimens, produces more than $2,000 per ton. It is said that $230,000 have been spent in working the mine, much of that sum having been produced by the mine itself. The mine was opened about 1853.

The Cahuabi mine, situated near the intersection of the hundred and twelfth meridian with the thirty-second degree of latitude, in the country of the Papazo Indians, produces a rich argentiferous copper ore. The mine was opened in 1859.

The Mowry mine, in the sierra of Santa Cruz, fourteen miles from the town of Santa Cruz, and 5,160 feet above the sea, has a rich argentiferous galena.

The Santa Rita mine lies in the sierra of Santa Rita, twelve miles east of Tubac. The ore is an argentiferous galena.

The San Pedro mine, east of the San Pedro river, produces argentiferous copper. The above are the chief mines in Arizona now wrought.

The Ajo copper mine, called, also the Arizona mine, one hundred and twenty miles south-east from Fort Yuma, is extremely rich. Its ores are oxides and sulphurets. The working of this mine proved unprofitable, and has been stopped.

Quicksilver Districts. § 88. California has some of the richest quicksilver mines in the world. The main quicksilver district lies in Santa Clara county, about sixty miles southward from San Francisco, and twelve miles southwestward from San José, in the coast mountains. There are three mines here; the New Almaden, which derives its name from the quicksilver mine of Almaden, in Spain; the Enriqueta, so styled in honor of Enriqueta Laurencel, little daughter of one of the proprietors of the mine at the time of the discovery; and the Guadalupe, a name suggested by the little river that drains the district. These three mines are all found within a distance of four miles; in one range of hills, nearly in a straight line with each other. The only one from which mercury is obtained, is a sulphuret or cinnabar, a red, heavy mineral, found, not in veins of regular width, but in large irregular masses, connected by small seams. The mines of Santa Clara county are at the junction of metamorphic limestone rock on one side, and eruptive rocks, chiefly trap, on the other. It has been frequently observed, that, at such points of junction, metallic ores are more abundant and rich than elsewhere.

"The mountain mass," says W. P. Blake, "in which the (New Almaden) mine lies, is of serpentine, with chloritic and talcose slates. Seams of limestone, intercalated in threads and masses of metamorphic limestone twelve feet thick, occur on the ascent before the serpentine is reached. The limestone is whitish, semi-crystalline, and without fossils. The trend is north-west and south-east, which is also the direction of the metalliferous veins. The dip is variable, but always to the east. Talc slate is the most abundant rock; but the serpentine and trap are associated with it in the mine. The gangue stone associated with the cinnabar is quartz forming geodic cavities. Sulphate of barytes occurs crystalized in some seams. The sulphuret of mercury is found in masses, towards which the vein lead."

The New Idria mine is in the coast mountains, about sixty miles south-eastward from San José.

Cinnabar is found on the sides of Mount St. Helena, in Napa county, and in the Geyser mountains, in Sonoma county. Several companies have commenced to open the veins in those districts; but it is not known yet whether the deposits will prove to have any value. One of the Napa companies has sent fifty pounds of metal to the market. A singular feature of the Napa and Sonoma cinnabar is, that the veins have much porous limestone rock, containing, in its interstices, liquid quicksilver, which flies out in minute globules whenever the rock is shaken violently.

Quicksilver mining is very uncertain, and requires a large capital. The irregularity of the deposits renders it impossible to know whether the mine contains much ore, or at what cost it can be taken out. Expensive furnaces must be built before the metal can be obtained from the ore. Many laborers must be employed in opening the mine and preparing for subsequent work.

The New Almaden mine has produced 3,000 flasks, seventy-five pounds in a flask, in a month; but 2,500 flasks may be set down as its present monthly yield. The Enriqueta mine has produced 1,000 flasks; but the average is much less. During the last quarter of 1860, the yield was about 2,400 flasks. The New Idria mine furnishes from three hundred to five hundred flasks per month, and the Guadalupe mine from one hundred and fifty to two hundred and fifty.

The ore of the New Almaden and Enriqueta mine contains about eighteen per cent. of metal; that of New Idria, about eight per cent.

New Almaden is supposed to be—next to the Spanish Almaden—the most valuable quicksilver mine in the world, and perhaps it is even superior to that. The mine is at an elevation of 1,000 feet above the sea, and two hundred feet below the top of the hill. Several hundred miners are employed, about half Cornishmen and half Mexicans, who are engaged in hunting the ore and taking it out. The deposits of ore have to be hunted; and the miners seek them by following up the little seams. Sometimes these masses are found fifty feet long, twenty wide and twenty high. The ore is hoisted to the surface by machinery, and then is hauled down to the *Hacienda* or Reducing Works, where there are fourteen furnaces of brick. Each furnace may be fifty feet long, twelve feet high and twelve wide. In front is the fire-place; next that, is a chamber for the ore, about ten feet cubic, with open walls on each side, so that the heat may enter from the fire, and pass into the condensing chamber behind, in which there are partitions, so that the smoke from the fire and vapor from the ore must pass up and down, alternately, half a dozen times, and finally it rises out of a chimney forty feet high. The ore is placed in the ore-chamber, in large pieces, and with open spaces between, so that the flames and smoke from the fire may pass through it. The earthy matter near the large deposits of cinnabar contains a good deal of metal, and is made into brick, so that they can be piled up, also, with open spaces for the fire to pass through. In the bottom of the condensing chamber is water, by which the fumes of the quicksilver are cooled and condensed. The sulphur of the cinnabar and the smoke of the fire escape through the chimney.

In the Enriqueta and Guadalupe mines, the quicksilver is collected in close iron retorts, which contain quicklime to absorb the sulphur.

The value of the New Almaden mine has been estimated very extravagantly, by the Attorney General of the United States, at $25,000,000; its real value is not more than one fifth that sum.

The quicksilver is put into wrought iron flasks, made of heavy sheet iron, about a foot long and five inches in diameter, with an iron screw for a cork at one end. Each flask holds seventy-five pounds of metal.

The New Almaden Company is now engaged in building a new and very large furnace, and in cutting a tunnel eight hundred feet below the present entrance of the mine. These improvements will enable them to increase their production considerably.

CHAPTER VI.

PROSPECTING.

Prospecting a River Bar. § 89. The prospector for gold should be familiar with the general principles of the distribution of gold, as explained in Chapter IV. Rich gold diggings are found only in districts where granite, quartz and slate exist together; but they may sometimes not appear on the surface of the ground, or even in the beds of the streams. If, however, a thorough geological examination proves that none of them approach the surface within fifteen or twenty miles, then it is useless to search for paying placers. All the rich auriferous districts are hilly or mountainous; have reddish earth, with numerous quartz pebbles and pieces of slate among the gravel.

The main implement used in prospecting is a pan made of stiff tin or sheet iron, with a flat bottom from ten to fourteen inches across, and sides from four to six inches high, rising at an angle varying from thirty to sixty degrees.

The prospector having found the district supposed to be auriferous, should go, in a season of low water, to some large stream where it emerges from a deep gorge, not far below which he will find a bar—a collection of sand. If the stream makes a turn on emerging from the gorge, there will be at high water an eddy on the side toward which it turns, and in that eddy the auriferous sand, if any, will be deposited. The prospector should go with shovel, pick, pan and knife, or spoon, to this bar, where he should dig down in a place where the sand is at least two feet deep, and as near to the water's edge as he can go, without having any water in his hole when he gets to the bed-rock. From the bottom he should fill his pan with dirt, taking care to scrape the dirt from the rock, and especially to clean any crevices that may be in the rock. When his pan is nearly full of dirt, he should take it to the water's edge, put it under the water, and then put his hand down into the dirt on all sides and raise it up, so as to have the water pervade the whole mass. If there be any pieces of clay, he should break them up and rub them in his hands until they are completely dissolved. Then taking hold of the pan on each side, keeping it under water, with the edge near him a little higher than the outer edge, he commences shaking the pan from side to side. The thick muddy water flows out, and its place is supplied by other water which finally carries away all the mud; but before the muddy water is gone, the light sand comes to the top, and flows out over the outer edge of the pan, which gradually gets a higher inclination and is raised out of the water, until at length only a few stones are left. The largest of these are scratched out with the fingers, and the shaking commences again, and presently all the earthy and stony particles are gone, and then the prospector may look for particles of gold. If he can find none on such a bar, he can say pretty safely that there is no gold in the basin of that river above the place where he has prospected. If the prospector have no one to show him how to use the pan, it might be well for him to practice beforehand, putting some rough little pieces of lead—not round shot, for they would roll out too easily—into the pan with some dirt, and when he can "pan out," so as to get rid of all the dirt and save every piece of lead, he will have skill enough for ordinary purposes. Men practiced in the use of the pan, sometimes give it a circular motion, so that the dirt runs round and round in it, thus dissolving the clay and throwing out the light material more rapidly than by a simple shaking from side to side. It is not difficult to learn both methods of "panning." The pan should be free from grease, the presence of which interferes greatly with all the processes for separating gold from earthy matter.

The gold is generally found imbedded in a stiff clay, mixed with gravel and stones. The object of the prospector is to dissolve the clay thoroughly and set the gold free, make a current of water to carry away the dissolved clay, and then to separate the metal from the stone by an agitation sufficient to throw out the lighter material, but not sufficient to throw out the gold. In panning, as in all methods of placer mining, the miner separates the gold from the dirt and stones chiefly by making use of the superior specific gravity of the metal.

Prospecting in a Ravine. § 90. If the dirt in the bar should prove rich, it is to be presumed that there are rich ravines not far distant. If the gold be very coarse, it may have come, not down the river, but from a ravine emptying into the river there. If the gold be fine, it probably came from ravines emptying into the river some distance above.

The best time to prospect gullies is during a rain or soon after it, when streams of water are found even in short channels. The prospector should find a place where a vein of slate, with strata nearly upright, crosses the gully. If such a vein can be discovered at a spot nearly level, but just below a steep part of the gully, so much the better, for the gold does not like to stop in precipitous places. The prospector now fills his pan with dirt from the bed-rock at the centre of the ravine, digging up some of the slate, if it be loose, and putting that into his pan too. He then makes a little dam in the ravine, and pans out in the standing water. I have said that the prospector should seek a vein of slate, with strata nearly upright, crossing the gully. If the strata be horizontal, they will present a smooth surface to the gold, which will slip over and go to some place below; whereas, if the strata be upright, the bed of the gully will be rough and full of crevices, in which the gold, when once lodged, will be safe against the fury of any current. Auriferous ravines and gullies differ greatly in wealth even though very near each other; and different parts of the same ravine differ also. The thorough examination of a large district of ravines, therefore, requires much labor. The reason why I recommend prospecting in ravines immediately after a rain is, because every little gully then has water for washing, and also because the dirt is easy to dig, and being filled with water is so much more easily washed.

Prospecting with a Knife. § 91. It might occur, however, that a person would wish to prospect in a very dry season, in a place without water for washing. In such case he should select a spot in a gully by the rules indicated in the last paragraph, dig away the earth to very near the rock, and then get down into the hole and scratch the earth over carefully with the point of a knife, picking out the particles of gold and throwing away everything else. He should be very particular to scrape out cleanly all the crevices in the rock, and if the rock be slate, he should dig up some of it slowly and carefully, examining all the seams for gold, which enters such places in a manner often very puzzling to the miner. In rich diggings, men not only prospect in this method, but work regularly at mining.

Prospecting a Flat. § 92. To prospect a flat, the miner should seek for signs of the place where the water ran before the earth was deposited; for every flat has a deposit of earth upon it, usually not less than six feet deep and sometimes as deep as a hundred feet. It frequently happens that the course of a brook on the surface of the flat indicates the position of an ancient brook lower down. The miner then should start in the bed of the surface brook and dig a hole or shaft to the bed-rock and try the dirt there. Nobody should undertake to prospect for deep diggings, whether in flats or hills, save in a district known to be auriferous. If gold cannot be found near the surface, there will be little encouragement for going deep. It is difficult to lay down rules in regard to prospecting for deep diggings. In many cases in California they have been discovered by accident. It has also freqently happened that miners at work in a little gully running from a flat or hill-side, have followed up a rich lead of placer gold until it took them into a class of diggings entirely different from that in which they started.

In an auriferous district where there are high hills of gravel, the miner should keep his eyes on them. Such hills often contain great deposits of golden wealth. The best places to examine these hills are where streams have cut down through them, exposing steep banks on each side. If distinct layers of clay and gravel be visible in the bluff, the prospector should try them all in his pan.

Prospecting for Quartz. § 93. Prospecting for quartz is entirely different from prospecting for placer diggings. Most of the auriferous quartz is found in veins running north-north-west and south-south-east, at an elevation from two to six thousand feet above the level of the sea. The prospector for quartz looks at every vein where it crops out on the side of a hill or in the bank of a stream, and if he cannot see any particles of gold, he usually infers that the rock is not auriferous. Rich quartz veins are often found by accident. In Tuolumne county, in October, 1858, a miner shot a grizzly bear on the side of a high and steep hill. The animal rolled down the hill until it came to a projecting rock, upon which it lodged. The miner went down to his game, and as he was skinning it, he saw gold in the rock, which proved, on examination, to be very rich. In May, 1855, a Mexican highwayman attacked a miner near Coulterville, Mariposa county, and after firing several shots on each side, the assailant was killed. The fight occurred in a ravine, and just after firing a shot which missed its aim, the miner saw a glitter of reflected sunlight from a rock where his ball struck. So soon as the highwayman was dead, he went to the spot struck by his ball, and there found a rich vein of gold-bearing quartz. The Allison vein, in Nevada county, reported to be the richest quartz mine in the State, was found by tracing up a rich lead of placer gold to the quartz from which it had come.

Mr. J. E. Clayton, a mining engineer, gave, through the columns of the *Mariposa Gazette*, the following advice to prospectors for auriferous quartz:

"The first step to be taken, is to ascertain the direction of the strata of the bed-rock and quartz veins imbedded therein. Then take a common pick, shovel, and good iron pan, and prospect the surface dirt along, and just under the break of the veins every few yards, thus following the vein as far as it shows itself, either by its outcrop or loose fragments; and if gold is found in the surface along the vein, it is good presumptive evidence that the vein is gold-bearing. Then ascertain the point on the vein that gives the best 'prospect,' and make a cut across it deep enough to show the vein as it is inclosed in the bed or wall rock; then make a careful examination of every part of the vein, so as to determine what part of it is gold-bearing. The casing of the vein where it joins the wall rock should be carefully tested also; it frequently occurs that the casing is richer than the vein itself. The best mode of testing the rock is to pound it up finely in a hand mortar, and wash it out in a pan or horn spoon. If a satisfactory result is obtained, then sink a shaft so as to cut the vein at the point where the best prospect is obtained, and follow it down, say forty or fifty feet. The character of the 'wall rock' should be closely observed, to ascertain the 'line of its texture.' The smooth faces that separate the vein from the wall rock should be carefully examined; the smooth faces have numerous small ridges and grooves upon them, that show the 'line of its projection,' or the direction from which the vein was forced up between the walls inclosing it. The ridges and fine grooves on the faces of the veins will, in most cases, be found to have the same direction of the texture of the wall rock; and the rich section of the vein will most generally continue rich in the 'line of its projection.' It is frequently the case that a vein will have a section of a few feet that will be rich, and all the balance of it be poor; therefore, it is very important to learn the 'line of its projection,' for the rich sections always follow the course indicated by the 'line of projection' and the 'line of texture' of the wall rock."

CHAPTER VII.

ASSAYING.

Kinds of Assays. § 94. The thorough miner ought to know how to make assays. If working in auriferous quartz or silver ore, he should frequently make assays to ascertain whether he loses any of the metal and how much, and to know whether the rock will pay for working, and how it should be worked, for the best manner of treatment will sometimes depend upon the richness of the mineral.

Assays are of two kinds, "qualitative" and "quantitative;" the former to ascertain whether a certain substance is in the mineral, the latter to determine how much. A qualitative assay of ore for silver is made to learn whether there is any of that metal in the ore; the quantitative assay shows the exact amount of the silver in it.

Means of Assaying. § 95. Assays are made with acids, by smelting in crucibles and by melting under the blowpipe. The processes are numerous and complicated, and some of them require a very nice knowledge of chemistry. I shall not attempt, therefore, to explain them all.

Some of the processes which I describe here under the head of "assaying," are often called "prospecting" by miners; but it is more convenient for me to treat of an examination of quartz rock made with a horn spoon as an "assay," rather than as a "prospect."

Gold Assay with a Spoon. § 96. Every quartz miner has a horn spoon for prospecting his lode and finding what part of the rock will pay. This horn spoon is made o an ox horn, one-half of which is cut away, leaving a bowl six or eight inches long and nearly three inches wide. He pulverizes his rock on a smooth, hard stone, a foot square. After breaking the quartz with a hammer, he uses a muller or hard smooth stone, about four inches square, to crush the quartz to a fine powder. He washes a handful of this powder in his spoon, which he uses like a pan, and if he can find a few specks of gold in a handful, he infers that it will pay. If he finds not a speck in a pound of rock, he infers that it will not pay.

Assay of a Metallic Substance. § 97. If an assay is to be made of a metallic substance to find out how much gold may be in it, a chip should be cut off from one corner, weighed in assayer's scales, put into a cupel, heated to melting, then withdrawn and allowed to cool. The cupel is a little cup made of bone dust for the special purpose of assaying, and when base metals are melted in it, it swallows them up, leaving the precious metals—gold and silver—pure. The button from the cupel is now melted with enough silver to weigh three times as much as the gold in it. This addition of silver is necessary to enable nitric acid to eat away the silver that was in the button, for when there is a little silver in much gold, the acid cannot get at the silver. There ought to be three times as much silver as gold to enable the acid to work to advantage. The gold and silver having been mixed in the proper proportions, are rolled out into a thin ribbon. This is boiled in nitric acid, which leaves the gold pure.

Gold Assay by Smelting. § 98. If auriferous quartz, free from sulphurets, is to be assayed, four hundred grains of the rock finely pulverized may be mixed with an equal weight litharge and five grains of charcoal. Put this mixture in a crucible large enough to contain twice as much more; then put the crucible in the furnace and melt the mass. Remove from the fire, allow the crucible to get cold, break it and the metal will be found in a "button" at the bottom, covered with a "slag" of melted rock and other matter. Treat the button as prescribed in the preceding paragraph.

If the rock to be assayed contains pyrites, it must be roasted till it ceases to give out sulphurous fumes. Mix four hundred grains of the powder with two hundred grains of litharge, two hundred grains of dry carbonate of soda, two hundred grains of dried borax and ten grains of charcoal; then put into the furnace and treat the button as directed in the last two paragraphs.

Presence of Copper Pyrites. § 99. Copper pyrites in quartz sometimes bears so close a resemblance to gold as to deceive even experienced miners; and, of course, it is far more likely to deceive the inexperienced. Indeed, iron pyrites often deceive these. The best method to discover the presence of pyrites, either of copper or iron, is to pulverize the mineral, put it into a saucer with some nitric acid, and that over a few embers, until dark red vapors rise. If pyrites be present, the acid will be discolored. Or resort may be had to the hammer; if the mineral flattens out on the anvil, it is gold; if it breaks into fragments, it is pyrites. The latter substance is usually in rectangular crystals; gold never takes that form.

Silver Assay with Testing Tube. § 100. The best qualitative silver assay for the general miner, is that with the testing tube. This is of thin glass, about five inches long and five-eighths of an inch in diameter; of a rounded bottom, with the same thickness as the sides. Enough of the mineral, finely pulverized, is put in to occupy an inch of the tube. On that is poured two inches of nitric acid. The tube is placed over a spirit lamp or a fire, till the acid boils. Nitric acid dissolves silver; and by this treatment, if there be any silver in the mineral, the acid must take it up. Filter the acid now through filtering paper, which can be had at the drug shops, and pour the acid back into the tube. Pour in a few drops of solution of common salt, and if there be any silver in the mineral, a white cloud or curd will be formed in the acid, by the silver precipitated by the salt. If there be no cloud, there can be no silver. If there be a cloud, the mineral contains either lead or silver. Pour off the acid, and expose the precipitate to the sunlight; in five minutes, if silver, it will turn purple; then pour on some spirits of ammonia, and the silver will be dissolved again. If a testing tube is not to be had, a common saucer may be used.

Silver Assay by Smelting. § 101. Silver ores are of two kinds; those containing lead, and those free from it. The former usually contain a large amount of lead, or rather, they are lead ores containing a little silver, and called argentiferous galena. The mode of assaying with the crucible, and also of working the lead-bearing silver ore, differs from that of the ore free from lead.

To assay argentiferous galena, mix four hundred grains of the pulverized ore with twelve hundred grains of carbonate of soda and forty grains of charcoal; put into a crucible, and that into a furnace; raise the heat sufficient to melt the mass; take out the crucible, give it a tap or two, to shake the metal to the bottom of the melted matter; let it cool; take out the button, which should be heated in a cupel as described in section ninety-seven, to drive off the lead, leaving the silver free.

Silver ore not containing lead may be assayed by mixing four hundred grains of ore with four hundred of litharge, eight grains of pulverized charcoal, two hundred grains of carbonate of soda. This mixture is put into a crucible, a thin layer of borax is sprinkled over it, and it is put into the furnace, and treated as directed in the preceding paragraph.

Assaying Gold Quartz by Weight. § 102. Phillips, in his little work on *Gold Mining and Assaying*, gives the following rule for ascertaining the amount of gold in a lump of auriferous quartz:

"The specific gravity of the gold—19,000.

"The specific gravity of the quartz— 2,600.

"These numbers are given here merely for convenience in explaining the rule; they do not accurately represent the specific gravities of all quartz and quartz gold. (The quartz gold of California has not, on an average, a specific gravity of more than 18,600.)

"1. Ascertain the specific gravity of the lump. Suppose it to be 8,067.

"2. Deduct the specific gravity of the lump from the specific gravity of the gold; the difference is the ratio of the quartz by volume: 19,000—8,067=10,933.

"3. Deduct the specific gravity of the quartz from the specific gravity of the lump; the difference is the ratio of the gold by volume: 8,067—2,600=5,467.

"4. Add these ratios together, and proceed by the rule of proportion. The product is the percentage of gold by bulk: 10,933+5,467=16,400. Then as 16,400 is to 5,467, so is 100 to 33,35.

"5. Multiply the percentage of gold by bulk by its specific gravity. The product is the ratio of the gold in the lump by weight : 33,35×19,00=633,65.

"6. Multiply the percentage of quartz by bulk, (which must be 66,65 since that of the gold is 33,35) by its specific gravity. The product is the ratio of the quartz in the lump by weight : 66,65×2,60=173,29.

"7. To find the percentage, add these two ratios together, and proceed by the rule of proportion : 633,65+173,29=806,94. Then, as 806,94 is to 633,65, so is 100 to 78,53. Hence, a lump of auriferous quartz, having a specific gravity of 8,067, contains 78,73 per cent. of gold, by weight."

To find the specific gravity of a lump of gold, quartz or auriferous quartz, divide the weight of the lump in air by the weight of an equal amount of water. To find the weight of an equal amount of water, deduct the weight of the lump in water from the weight of the lump in air. When the lump is to be weighed in water, it should be suspended by a horse-hair so as to hang into the water; keeping, of course, all other parts of the scales clear of the water.

Importance of fair samples in Assaying. § 103. In making assays of auriferous minerals, as, indeed, of all ores, the first point is, to get a fair sample of the mineral; and whenever the result of an assay is to be considered as the basis for the purchase of the claim, or the investment of money, the person proposing to purchase or invest should first satisfy himself that the sample assayed was a fair average specimen. He should know that the sample was chosen with all the honesty and precaution of which such cases admit. Metalliferous vein stones vary greatly in richness; in some places presenting nearly pure metal, in others, being almost barren. It is a very common occurrence in mining countries, that dishonest men select rich bits of ore, have them assayed, show the assayers' certificates, assert that the samples assayed were fair samples of the vein, and try to find purchasers on the credit of their assertions. The proper way to guard against such frauds is, to know that the assay fairly represents a considerable quantity of mineral, which, itself, fairly represents the body of the claim as near as possible. Only very small amounts can be assayed; but it is an easy matter to ascertain, by assay, the value of a large quantity of rock. A ton of rock may be pulverized finely, then mixed up together carefully, and the powder will be of the same quality throughout; and then an assay of an ounce of it will truly represent the whole ton; whereas, if an assay had been made of an equal weight of rock broken off at random, from the vein, the result would not have given any trustworthy indication of the worth of the rock. In case that dust is to be assayed, it should be melted and stirred together, and then chips chiseled off from opposite corners may be assayed. It has frequently happened, for instance, that gold miners in California have brought their dust to San Francisco and deposited part of it in the Mint, and part of it in a private refinery, and found that the deposit in one place was estimated at a higher value per ounce than in the other; and this with dust, all of which had come from the same claim; and then they have made charges, or had suspicions of dishonesty. But this was a mere suspicion, without good cause, unless the discrepancy was larger than any I have ever heard of. In such case, if the miner had wished to test the accuracy of the assays, he should have made sure that the samples to be assayed were of precisely the same quality, and that can only be done by melting together, if the substance be metal, or mixing together in powder, if it be an ore.

CHAPTER VIII.

MODES OF PLACER MINING.

List of Modes. § 104. The modes of placer mining are numerous, and most of them are named after the instruments used. The principal are knife mining, dry digging, dry washing, panning, mining with the cradle, with the quicksilver machine, the tom, the sluice—of which last there are several kinds—and the hydraulic process.

Knife Mining. § 105. Mining with the knife is the simplest mode of obtaining gold. It can be pursued with profit, as a business only, by experienced miners, in diggings rich in coarse gold. The knife miner must know where to look for the richest spots, and avoid everything else. He seeks for crevices, from which he scrapes all the dirt, picking out the separate pieces of gold, if it be coarse, or if it be fine, putting gold and dirt together into his pan, and panning out when his pan is full, or when he has done with a crevice.

Dry Digging. § 106. "Dry digging" differs from knife mining in this, that the latter requires the use of the knife and pan, whereas the former may require the use of the pick and shovel to strip off the top dirt, and does not require the pan. "Dry digging" is a mode of mining, and is not to be confounded with "dry diggings," a kind of mining ground. The process of dry digging has been described in section ninety-one, under the head of prospecting; for prospecting with the knife is the same as dry digging.

Dry Washing. § 107. Dry washing is used in very rich coarse gold diggings, where there is no water. The miner tosses the dirt into the air while the wind is blowing, and thus gradually winnows out the gold. The Mexicans have done more work at this kind of mining, in California, than any other class. The Mariposa *Gazette* thus describes the process as pursued by them in that county :

"During the dry summer months, the Mexican miner may be seen, at almost any hour during the day, coyoting (burrowing like the coyote, or small Californian wolf) for gold in the neighboring hills or the adjacent flats. Sinking a square hole, some four or five feet deep, to the bed-rock, he carefully scrapes all the dirt lying immediately on the ledge into a wooden *batea*, (or pan) which he carries to the nearest tree, and under its shade pounds up the hard lumps of earth, until nothing but dust remains. A bullock's hide is now spread out upon a level spot, when the Mexican raises the *batea* above his head, and with an oscillating motion shakes out the dust upon the skin, until all the dust has fallen. This process is renewed for a number of times, until very little of the original mass remains, which is carefully collected and placed in a pile separate from the unpounded earth. When it is found that the claim from which the dirt has been taken pays rich, or even reasonably well, the Mexican returns to his diggings, and commences to cut into the sides of his hole, just above or adjacent to the bed-rock. They are a species of badger miner. Sticking close to the ledge, they will burrow with their light crow-bars for a distance of six or eight feet, ascending or descending with the ledge, following it closely, and carefully scraping up the earth upon its surface. They seldom use any other tools except the small crow-bar, which is pointed at both ends, the *batea* and the horn spoon, with which they scrape and rake up the soil, after first loosening it with the bar. They are by no means selfish in their mining operations. When one strikes a good claim, his neighbors and friends are soon informed of it; but it is only to their own countrymen to whom he is thus disinterestedly generous. When one claim has proved good, the whole of the gulch, flat or hill, is soon taken up by his compatriots, and then begins the work of coyoting, in which they seem to delight, and which gives so remarkable an appearance to the mines wherever they have been working. Dry washing requires considerable slight of hand in working to advantage. A windy day is preferable for this manner of washing, as the wind more rapidly carries off the fine dirt, while the great density of the gold removes all fear of its being carried off the hide, even by the strongest breeze. The Mexicans make a good living during the summer months at dry washing, and in many instances we have known them to realize small fortunes by this manner of washing."

Panning. § 108. The process of panning has been described in section eighty-nine, under the head of prospecting.

It sometimes happens, in mining with the pan, but much more frequently in mining with the rocker, that a large quantity of black sand, full of fine particles of gold, is collected. The black sand is very heavy, and cannot be separated from the fine gold by panning; and blowing must be resorted to. This is done in a "blower" of tin or brass, a dish from four to ten inches wide, and twice as long as broad, open at one end, with a rim an inch high, at the other end, and the two sides. Into this blower the black sand and gold are poured; and while the mouth of the blower is raised a little above the level, the miner blows the sand away, gently, with his breath, occasionally shaking the blower, so as to change the position of the particles.

The Rocker. § 109. The cradle or rocker is, next to the pan, the most simple instrument for washing gold. It resembles, in size and shape, a child's cradle, has similar rockers, and is rocked in a similar manner; whence its name. The cradle box is a wooden trough, about twenty inches wide and forty long, with sides four inches high. The lower end is left open. On the upper end sits a hopper or riddle, which is a box twenty inches square, with wooden sides four inches high, and a bottom of sheet iron or zinc, pierced with numerous holes, half an inch in diameter. Under the hopper is an apron of wood or cloth, which slopes down from the lower end of the hopper to the upper end of the cradle box. A strip of wood an inch square, called a riffle bar, is nailed across the bottom of the cradle box, about its middle, and another at its lower end. Under the bottom of the cradle box are nailed two rockers, so that a rocking motion may be given to the machine. Sometimes an iron spike runs down from the center of each rocker, and enters a hole in the bar of wood on which the rocker rests. The purpose of the spike is to keep the rocker from moving sidewise, or slipping downwards. The wooden bars on which the rockers rest may be connected together by cross pieces, so as to form a square frame.

When the rocker is to be used, it is placed in the spot to which the pay-dirt, and a constant supply of water, can most conveniently be brought. The lower end of the cradle is placed so as to be about two inches lower than the upper end. The miner fills his hopper with pay-dirt, sits down by the side of his cradle, pours a dipperful of water upon the dirt, and begins to rock, and keeps on pouring water and rocking until nothing remains in the hopper save clean stones. He then rises, lifts up his hopper, throws out the stones, and is ready to repeat the operation. It very rarely happens that he finds pay-dirt, all of which will pass through his riddle. The length of time required for washing a hopper of dirt depends upon the tenacity of the dirt, the supply of water, and the violence of the rocking. If the clay be very tough, a quarter of an hour may be spent in washing a hopper; if it contain much sand, two or three minutes may be enough. The water, dissolved clay, sand and gravel and gold, less than half an inch in diameter, fall down through the holes upon the apron, which carries them to the upper end of the cradle box, whence they run down towards the open end. The gold and heavier particles of gravel are caught behind the riffle bars; the water, thin mud and other light materials are carried away over the riffle bars, and are considered worthless.

The rocker requires a large supply of water, which should be supplied by a little brook, with a reservoir large enough to receive the dipper, and near enough to the miner to enable him to reach the water without moving from his seat by his cradle. Both the water and the rocking are necessary to wash with the cradle; both are needed to dissolve the clay and carry away the light and soluble matter, while retaining the gold. The rocking would do no good without the water, and the water would do little good without the rocking. As almost a constant stream of water pours into the hopper from the dipper, so almost a constant stream pours out at the lower end of the cradle box.

If the gold is very fine, the hopper may be put over the lower end of the cradle, so that the apron may be longer, and much of the gold is then caught on the apron.

The rocker must not be set level, for in that case too much

dirt would accumulate above the riffle bars, and would "pack" or settle down into a hard mass, on a level with the riffle bars, and all the dirt and gold coming down after it had once packed, would run away as over a smooth board. If, on the other hand, the inclination of the rocker be too steep, the current of water is too strong and carries away the gold with the dirt.

Packing is a serious difficulty in the way of all or nearly all processes of gold washing. The dirt will pack in cradles and sluices; and when it has once packed, there is little obstacle to the escape of the precious metal. Many devices have been used to prevent packing; but I never knew one to succeed. Sometimes, the bottom of the cradle is made of sheet iron, and of a concave shape, being about two inches deeper in the middle than at the sides; but the dirt packs in these cradles nearly as badly as in the others. If I had need of a rocker now, I think I should try one with a convex sheet iron bottom, the convex side up, with a riffle bar considerably higher at the sides than in the middle. Quicksilver has been used in cradles, to prevent packing and to catch the fine gold; but in most cases some of the amalgam is lost, carrying away gold that would otherwise be saved, and it does not prevent packing. The more constant the rocking of the cradle, the less the danger of packing. A device to prevent packing, is to put a little block under the rockers at each end, so that every time they come down the cradle gets a jolt, shaking up the gravel on the bottom and letting the water get under it, and thus preventing its settling. A rocker always furnishes work for at least two men, and the dirt does not pack so badly when two are at work as when there is only one; for in the latter case, after washing a hopper, he must always move from his seat, take up his shovel and fill his hopper, and then go back; whereas, if there are two, the "shoveler" can fill the hopper as soon as the "cradler" has emptied it. The cradler has a large iron spoon, with which he occasionally scrapes over and loosens the dirt that has lodged above the riffle bars, and several times in the course of the day, he "cleans up," by taking out the dirt into his pan and panning it out. The upper riffle bar always catches much more gold than the lower one; and sometimes cradles are made about two feet long with a single riffle bar. These are made only when they are to be frequently moved.

The cradle should be placed, if possible, so near the claim

that the pay-dirt may be shoveled directly into the hopper; but a greater weight of water is required than of dirt, and if water cannot be brought to the claim, the dirt must be taken to the water. The mode of carrying the dirt depends upon the distance and the nature of the road. If the distance be small, men carry the dirt in buckets, or wheel it in wheel-barrows; if great, pack mules, carts or wagons are used. When the water can be brought to the claim, two men are usually enough, in shallow diggings, for one rocker. But if six or eight feet or more of barren dirt is to be " stripped " off, before reaching the pay-dirt, three or more may be required. Sometimes a laborer is occupied with bailing water out of the claim, and attending to the " tailings," as the gravel and sand which escape at the lower end of the rocker are called. These " tailings" are deposited by the water after leaving the rocker, and soon accumulate in a formidable amount, if not carried away by a swift descent.

A miner alone should wash, in ordinary shallow diggings, from seventy-five to one hundred and fifty " pans " of dirt in a day with a cradle; and two, twice as many. A pan is an indefinite amount, varying from half a peck to a peck; perhaps, usually, about half a cubic foot of dirt. Frequently the shoveler has a pan or bucket, which he fills with his shovel, and when the cradler is ready for him, he picks up the pan and empties it into the hopper.

In some pay-dirt the clay is so tough that more than an hour would be required to completely dissolve a hopperful of it. Sometimes the cradler undertakes to mash up the lumps with his hands; sometimes he rocks his cradle and pours his water for five or ten minutes or so at a hopperful, and throws out all the lumps undissolved at the end of that time, intending to wash them over again after they shall have been softened by exposure to the weather. Sometimes the dirt is dug up and exposed to the weather before it is washed.

The Puddling Box. § 110. Another device for dissolving tough auriferous clay, is the puddling box. This is a rough wooden box, a foot deep and five feet square. The clay is thrown in with water, and worked about with a hoe until dissolved, when a peg is taken out of an auger hole about four inches from the bottom, and the thin mud or " slum " runs out, leaving the heavier material at the bottom. The work continues in this way all day, and at night the contents of the box are taken out and washed with a cradle or pan.

The Long Tom. § 111. The tom or long tom is a wooden trough, from eight to fourteen feet long, eight inches deep, usually about sixteen inches wide at the upper, and thirty inches wide at the lower end. The bottom at the lower end is made of a riddle or perforated sheet of iron, and under this riddle is placed a riffle box, or small trough with several riffle bars. A constant stream of water runs through the tom, entering at the upper end, where the dirt is thrown in. The riddle has an upward turn at its lower end, so that nothing can run over there. The large stones are thrown out with a shovel, and the small ones escape with the sand and gravel through the riddle. The gold is all caught in the riffle box, where the dirt is kept loose by the water falling from above. Sometimes quicksilver is put into the riffle box to catch the fine gold. From three to six men may work with a tom. The tom is better suited for level ground than the sluice, which requires a considerable descent for the water. The tom is very seldom seen now in California.

The Quicksilver Machine. § 112. The quicksilver machine or Burke rocker, is a long cradle on stilts, with reservoirs of quicksilver in the bottom. It is about seven feet long, two feet wide and two feet high. An immovable riddle or perforated plate of iron forms the top of the machine throughout its length. Under this is the box containing a number of riffle bars, and above each one some quicksilver is placed. The dirt is thrown upon the head of the riddle, where a stream of water plays constantly through a hose, and the rocking motion of the machine and its downward inclination keep the dirt moving gradually toward the lower end, where the stones are allowed to escape, but the lumps of earth not dissolved are pushed back under the water and retained until they disappear. The quicksilver machine requires at least four men to work it, and in many places seven or eight men are necessary. It is suited only for fine gold, for if the gold be coarse it might be caught with far less trouble in the cradle, tom or sluice. The machine is cleaned once a day. All the gold is found in the mercury, which is squeezed through buckskin and the amalgam retorted. Quicksilver machines are great rarities in the mines now, though pretty extensively used previous to 1852.

The Board Sluice. § 113. The board sluice is the most important of all mining inventions for washing dirt. It is a large wooden trough, from one to five feet wide, and from fifty to fifteen hundred feet long, or even longer—the longer the better. It has numerous riffle bars, and an inclination varying from an inch to an inch and a half in a foot. The larger and longer the sluice and the greater the amount of water, the steeper the inclination.

The sluice is made of inch and a half boards, twelve or fourteen (usually twelve) feet long, and in sections of that length. These sections or boxes are three inches wider at the top than at the bottom, so that they fit into each other. They can thus be put together, taken apart and hauled about with very little trouble. The boxes stand upon trestles, two or three under each box. Very rarely does the sluice lie its whole length on the ground. The inclination of a sluice is called its "grade." If there is a descent of twelve inches in each box of twelve feet—the descent being usually uniform throughout the sluice—it is said to have a "twelve inch grade;" or if the descent be eighteen inches, then the sluice has an "eighteen inch grade." The depth of the sluice box is from one-third to one-half of its width. Sometimes the sluice is made double, with a longitudinal division through the middle. The advantages of this plan are, that it may be used by two companies or one, that it can be used with a large or small supply of water, and that while "cleaning up" is in progress on one side, the ordinary washing may continue on the other.

A vast amount of dirt may be washed in a sluice. The largest size, four feet wide, will wash a mass twenty feet cubic of dirt in a day, equal to four hundred and fifty cubic yards; but this amount of dirt can only be supplied by a hydraulic. A small sluice will wash all the dirt that can by thrown in by from five to fifteen men. One man is often required to see that the sluice does not choke—that is, that large stones and lumps of clay do not collect in one spot to dam up the water and drive it over the sides. In small sluices, a "sluice-fork" is sometimes used for throwing out the large stones. This fork is one invented for this special purpose. It has five tines, three inches apart, about a foot long; blunt, and of equal width from heel to point. The tines are made blunt so that they may not catch in the wood, and that stones may not get wedged in between them.

A constant stream of water enters the head of the sluice, and runs through its entire length. The size of the stream varies from twelve to two hundred inches. (See ¿ 125.) When the sluice is used to wash the dirt of a hydraulic claim, the amount of water is very rarely less than forty inches. From eighteen to twenty inches are, however, considered the usual supply for a board sluice not connected with a hydraulic claim, and that amount is called a "sluice-head."

In the bottom of the sluice are placed longitudinal riffle bars, which are six feet long, from two to four inches wide, and from three to seven inches high. They are put down an inch or an inch and a half apart, and are wedged in their places. There are two sets in a box, the riffle bars being only six feet long while the boxes are twelve. In rare cases the riffle bars cross the box diagonally, running downwards from one side, then from the other.

The great body of water rushing down through the sluice, hurrying with it many large stones, rapidly wears out the sluice boxes, or wood in them exposed to the friction. In hydraulic claims, all the stones run through the sluices, some of them weighing one hundred and fifty or two hundred pounds. Larger boulders are broken up with hammers, and reduced to a size which may be safely allowed to enter the sluice box. The sides of the sluice boxes are protected by boards, which must be renewed frequently. The riffle bars suffer most, and in hydraulic sluices must be renewed every week. A plan has lately been devised, however, to make "block riffle bars," sawn across the grain, and only two feet long. When fastened down in the sluice the grain will be perpendicular, and the wood will not be worn away so rapidly as when the grain lies lengthwise in the box.

The spaces between the riffle bars soon fill up with stones and dirt, but there are such irregularities in the surface that there are numerous little cavities where the particles of gold, quicksilver and amalgam will be arrested. In a couple of hours after washing has commenced, some quicksilver is put into the sluice at the head, and it gradually works its way downward, catching gold as it passes along. When the riffle bars are placed diagonally in the sluice, they do not touch the side at their lower ends, but leave an open space through which stones and quicksilver can pass, and going through they strike the next bar, which carries them to the other side, and so they go rolling from one side to the other. A vessel of quicksilver with a small hole in the side, so as to allow the liquid metal to escape in drops, stands at the head of the sluice, and these drops run their zig-zag course down the sluice, overtaking all the gold and catching some of it, and being themselves caught in longitudinal riffles near the end. These diagonal riffle bars are, however, very rarely used, and only in small sluices.

The period of time from the commencing to wash in a sluice to the cleaning up, is called a "run." In very large sluices, a run lasts till the riffle bars are worn out—usually in six or eight days, which is the ordinary duration of a run in all classes of board sluices. Many sluice miners clean up on Sunday; it is light work, and they have got into the custom.

Cleaning up commences with taking up five or six sets of riffle bars at the head of the sluice. Most of the gold and amalgam that was caught in these riffles, now lodges above the first set left in the box. A man with a scoop and a pan takes up this precious material; then five or six more sets of riffle bars are taken up, and so on.

From five to twenty men can work with a sluice. Most of the work is necessary to dig the dirt, and as this is done by the force of water in hydraulic claims, fewer men are required in hydraulic sluices than in others.

Mr. A. B. Paul, an authority in these matters, gives the following advice to sluice-miners as to the best method of saving very fine gold:

"Get a sheet of copper-plate, say three feet long, but as much more as you like, and eighteen inches wide, or whatever width you may desire your sluice boxes. Also, get a sheet of iron-plate—or very heavy Russia iron might do. You want it strong enough to stand the wear and tear of sluicing, and of the same length and breadth as your copper-plate. Have this iron perforated by slits half an inch in length, and not over a sixteenth of an inch wide, and not have the openings follow each other in a row, but change position on every half inch. So any fine substance floating over, is bound to go through some one of them. You will of course have the length of the openings to run with the width of the plate. These can be got up best by those accustomed to making screens for quartz mills.

"You now want some quicksilver, say ten pounds, and a pound of nitric acid, and we will then 'go to work.' You have all the material for working, excepting the sluice boxes. On opening the copper-plate for use, and which for convenience in packing you have probably rolled up, you will see that it is hammered out and lies perfectly flat. Arrange your sluice boxes at whatever grade is best suited to the dirt to be washed, and of what length you want, as I have only to do with the last one. All being set, we will now line the bottom of the last box with the copper-plate, which, as I said before, you want to have as level as possible. Inside, and on each side, nail a strip of board, say six inches in width. This will hold the coppers in place, and keep the amalgam from working under the edges; besides, they act as the support for your own perforated plating, which now is set on the strips, and directly over the copper plating.

"We will now take up the iron plating, which for convenience should be made movable, and prepare the coppers for use. Then set the box; take a small portion of your nitric acid; mix it half and half with water; then take a rag, or sponge, or whatever may come handy, and wash the exposed surface of the copper. Having done this, take a little of your silver, drop it on, and rub your plate thoroughly till it is all perfectly silvered. Now set your sluice at the same grade as the others, but drop it so as to bring the iron plating on a level with the bottom of the one above. It will be seen at a glance that the larger and heavier material rushing down the sluices will easily glide off, and allow the finer particles of gold— which by this time have gained the bottom—and other fine substances to descend, with a portion of water, through the openings on to the coppers below. If possible, let a small stream of fresh water in at the head of the coppers, but be careful not to have too much. This will thin the material coming through, and allow the gold to be more readily caught. You want a riffle below your coppers, to catch whatever quicksilver may run off of them by overcharging. Your coppers at first will turn green, but no matter; every morning, for the first few days, rub it off, and put more silver on. As gold collects, this green will disappear.

"It is best not to touch the plates until you are through your working, unless, as I said before, they become dirty—*too* heavily charged, and the gold is unsafe to leave. It is an attested fact, well understood by all workers in gold, that nothing catches it and returns it better than amalgam."

6

Sometimes transverse blocks of wood are used for riffles. They are cut across the grain, from two to four inches deep along the grain, and as wide as the sluice. These blocks are wedged into the sluice boxes, with transverse spaces of an inch or two between them.

Another device is, to fill the pores of such blocks with quicksilver. This is done with an iron cylinder, with a sharp edge, which is driven into the block a little way and then the quicksilver is forced down through the cylinder into the wood.

Some dirt (called cement) is so tough that it cannot be dissolved by running once through a sluice, nor even by running through twice; so they save the tailings, and after leaving them exposed to the air for a while, wash them—in all, three times. The third time completely dissolves the hardest dirt.

The Rock Sluice. ₰ 114. The "rock sluice," or cobblestone bottom, is the best of all in places where it can be used to advantage. It wants a steep grade, a large body of water, and a wide sluice box. Mr. B. P. Avery wrote thus of the rock sluice for the San Juan *Press:*

"One of the latest improvements in mining is the introduction of the rock sluice. The hydraulic power, well directed, tears down and washes off the auriferous earth with all the power and effect of natural forces directed by reason; while the blocks and riffles lining the sluice boxes, through which the dissolved dirt is conveyed, are only cunning substitutes for the gravel beds of natural water courses. These same gravel beds are now more closely imitated by lining the bottoms with cobble-stones, lapped one over another in regular layers, and inclining down stream. This idea was crudely adopted several years ago, rocks being piled irregularly in the sluices, and there allowed to remain for an indefinite time. The plan, now, is so systematized as to be really valuable. Every section of sluice, or each box fourteen feet long, is regularly paved as above described, the stones held firmly down by nailing strips of board, five and a half inches wide, on each side of the box, and wedging a cross-piece under these strips at the end of each box. As soon as the dirt and water have been allowed to flow over the gravel bottom, it becomes immovable, as though set in mortar. The paving can be rapidly accomplished, one man being able to finish, in a day, twenty-five boxes, fourteen feet long and thirty inches wide each. The material lies at hand in nearly every mining claim, and costs nothing but the labor of appropriation and selection.

"The advantages of rock sluices may be briefly stated. Those who have had long expceience with them, assert positively that they save more gold than any other sluices in use, and a kind of gold which no other sluices save at all. Mr. Welch, of Indian Hill, Sierra county, who has 2,300 feet of rock sluice leading from his claims, declares that he saves twenty per cent. more gold than he ever did before, out of the same dirt. He has thoroughly tested the matter by having alternate sections of rock and block sluice, and invariably obtained most gold from the former. He, as well as others, has observed that the rock sluices save the most fine gold, the almost impalpable powder of the precious metal, which is generally lost. For the same reason that more gold is saved, less quicksilver is lost.

"The rock sluices also effect a great economy of lumber. All other sluices are lined with blocks of wood about three inches thick, the cost of which, for each section fourteen feet long and thirty inches wide, is four or five dollars. These blocks have to be frequently renewed, owing to the great friction of rocks, earth and water running over them. In some instances, they will not outlast twenty days of washing. This was the case in the claims of Mr. Welch, where the saving effected by discarding blocks amounts to a very large sum. In his 2,300 feet of sluice there are, say one hundred and sixty-four boxes, that would require new blocks every twenty days; in three hundred days, each box would cost, at four dollars for every new lining, sixty dollars; and the expense of the entire sluice for the same period would be $9,840. In the Kentucky claims, at Sweetland—where may be seen a very handsome specimen of rock sluice—the saving on the blocks for sixteen boxes, at four dollars each, amounts to about sixty-four dollars every forty days that washing is done. Here, then, without reference to the superiority of rock bottoms as a gold-saver, is effected an economy that would alone render many unprofitable claims sources of income to their owners. One more recommendation of rock sluices is found in the fact that they offer fewer facilities for robbery. Thieves can help themselves in block sluices by simply scooping up the amalgam, as it lies in narrow crevices between the blocks; but here it is buried in sand, among stones hard to remove, and needing to be washed.

" Rock sluices are constructed upon a grade of from fourteen to sixteen inches for every fourteen feet, the heaviest dirt, or that which flows with least freedom, requiring the most grade. They cannot, ordinarily, be laid through tunnels, because these have to be run on as light a grade as possible—say one inch to the foot—and block sluices are used in them, as offering the least impediment to the flow of rocks and dirt. Their prime value is to receive the 'tailings' at the mouth of tunnels, and convey them for long distances down hill-sides. The boxes are, usually, thirty inches deep, and thirty inches wide, a greater width being obtained sometimes by constructing parallel lengths with a low partition. Flat, oval-shaped rocks, the size of a man's hand, only thicker, and as hard as possible, are selected for the bottoms. When the miner wishes to clean up, say after washing ten or twenty days, the stones are loosened with a pick, washed off by allowing ten or twenty inches of water to flow through them, and then laid out until the boxes are washed down and cleaned of their golden gatherings. The whole process is simple, economical, and worthy of trial by every miner."

The Tail-Sluice. § 115. The " tail-sluice " is a sluice to wash tailings. It is a very large sluice, is almost invariably paved with cobble-stones, and is allowed to run for months without cleaning up. It is put down in the bed of a creek, and the owners do little or no work at it from the commencement of the run until they clean up. Some of the most notable tail-sluices in the State are at Sweetland, Nevada. Mr. Avery thus wrote of one at that place :

"'The flume is laid along the bed of the creek, below the town of Sweetland. It is constructed of inch and a half and inch and a quarter lumber ; consists of two parallel sets of boxes, which are laid snugly side by side, and each of which is four and a half feet wide by two and a half deep, making a total width of nine feet ; supported on heavy posts and stringers, and banked in solidly on either side by gravel, which has been allowed to rise to a level with the flume, for the purpose of anchoring it, by means of dams here and there. The total length of flume is about 1,500 feet, 600 feet of which are laid through a tunnel which pierces a point of land, thus saving a considerable distance, and the total cost is stated at from $12,000 to $14,000. At one place, a point of land has been cut down and washed off, revealing indisputable evidences of a slide at some remote period. Large cedars were found entire, under the mass of rock and earth, and on being chopped, proved to be in a tolerably sound condition, though in the first stages of that transformation which converts them, under the influence, doubtless, of sulphurous acid and iron, into a black mass, resembling charcoal, and which curls up in dry, smooth chips on being exposed to the air. The face of the excavation is veined with oxide of iron, which colors the soil except toward the bed-rock—consisting of slate—where it is of a bluish lead color, heavily impregnated with sulphuret of iron, and giving off, on exposure, the peculiar offensive odor of sulphureted hydrogen.

"On the surface of this slide are growing larger trees than are buried beneath it. The fallen mass was evidently once supported on the brow of the ridge above by a projecting cliff of slate and mica schist, huge fragments of which are seen in the slide, and lying above the buried cedars. But we are digressing from the subject of the flume. The last section of it is laid at the base of a perpendicular cliff of rock, close to the river bank. This cliff is at least one hundred and twenty-five feet high, and its summit has been worn in twenty feet by the action of the water, leaving a half circle, the two points of which hold in their grasp an immense boulder, behind which, as into a huge vessel, leaps the roaring cataract of chocolate-colored water and stones, sending forth its muddy spray and icy breath as it leaps, and at one plunge striking the worn and slimy granite below.

"Sweetland Creek receives the tailings of a large number of claims. These tailings lie along its channel for a distance of more than a mile, and are fifty feet deep in places, having half swallowed up the trees which stand in the bottom, and deprived them of their vitality."

The Ground Sluice. § 116. The "ground sluice" is not made of boards, but is a ditch cut in the mining ground. It is used where the dirt is too poor to pay for washing with a board sluice, and sometimes because the surface is too steep to use a board sluice conveniently. The ground sluice can only be applied where the water can have a rapid descent. Sometimes the ditch is cut to the bed-rock before turning in the water; sometimes it does not go to the bed-rock at all; sometimes the water is used to wash away the dirt down to the rock. The water having started, the miners commence to throw in the dirt. The water is in a deep gully, and the miners pry off great masses of earth from the banks, and let them fall down into the ditch. A sluice fork is used to remove stones from the bed of the ground sluice, but some stones are always left to catch the gold. A run in a ground sluice lasts for several weeks or months. A board sluice or a tom is used for cleaning up. The ground sluice or its principle has been applied for various purposes. Part of the cutting of the Sacramento Valley Railroad was done with a sluice, and the ground, being auriferous, paid for the labor. In Forest City, the deep snows which fall there are sluiced out of the streets, and in Placerville they have cleaned the streets of mud on the sluice principle.

The Sluice Tunnel. ₂ 117. The name of "tunnel sluice" has sometimes been applied to a tunnel cut to be used as a sluice, but the more proper name is a "sluice tunnel." The following is a newspaper description of a sluice tunnel in Yuba county:

"North of Camptonville, about three-fourths of a mile, there is a basin formed by a circle of hills, which yields, during the continuance of water, above average wages. When this was discovered, it was apparent that mining could not be prosecuted there unless there was some outlet found or formed through which tailings could be carried off. On learning this, a company of eight miners, called the Ravine Company, offered to push a tunnel through the hill, so that the tailings might be washed directly down to Oregon creek, which runs at the foot of the outside hills that form the basin, provided the tailings should be given them for the work. This being agreed to, work was commenced in May, 1855, and continued until January 1st, 1857, when the last blast was fired and the tunnel completed. The tunnel proper is run through solid bed-rock; that is, until exposed to the air and sun, almost as hard as flint, and is 1,250 feet long, five feet wide and six feet high, and has a fall of forty-three feet. The expense of running it was something over $20,000. The tunnel and branches that were run into the hill for the purpose of prospecting the ground, make in length one and a quarter miles, and all in the bed-rock. Fifteen companies, with an average of twenty-five pipes, making eight hundred inches of water per day, run their tailings through this sluice, which yields to the Ravine Company an average of $1,200 per week for six or seven months in the year. Besides the tunnel, the company own thirty-five claims of seventy-five feet square on the hill. These are worked on the hydraulic principle, but it is difficult to judge what the claims pay. It is probable that it will take from five to eight years to work out the gravel in these hills, and fifteen years to work the gold entirely out, as the bed-rock will be washed up to the depth of four or six feet as wages decrease. The bed-rock, on exposure to the sun, becomes almost as soft as clay, notwithstanding its extreme hardness in the bed, and would pay about two dollars per day by washing the crust or upper layer, but on descending it becomes poorer, until at last it is worthless. The Ravine Company's tunnel is one of the finest in that section of the country."

The Under-Current Sluice. § 118. The "under-current sluice" is another variation from the simple sluice. Mr. R. Dunning, who claims to be the inventor of it, gives the following description of it and exposition of its merits:

"By means of two or more iron bars at the termination of a section of sluice boxes, forming a right-angle grating, a portion of the dissolved earth, fine gravel and water is separated from the lumps of hard earth, cobble stones and gravel, and drops into a set of more gently graded sluice boxes beneath, when they flow slowly off in another direction, while the body of water and coarse material dashes down a 'dump' or 'fall,' to be again taken up in sluices with the tailings from the under-current, and subjected anew to separation.

"This process insures a more thorough amalgamation and saving of the particles of gold, the most of which drop through the grating into the under-current, where, being subject to a less violent motion, and passing through a greater variety of riffles, they are more likely to be finally arrested. It effects a large saving of rusty gold, which will not readily amalgamate. It gives more opportunity for saving gold in a short distance, and to scour cement without loss of tailings and grade.

"On hill-sides, where there is abundance of space, it is a valuable adjunct to tail-sluices: where the latter terminate at the river's edge, and would otherwise discharge all of their contents into the stream, the under-current can be made to receive the best portion of the tailings, and convey them for any distance along the bank.

"The immense friction of rolling rocks being removed, the under-current effects a saving in false bottoms of about seventy-five per cent.

"At San Juan Hill, Nevada county, where this invention was first introduced, and is now extensively used, it is considered a valuable improvement—saving both gold and quicksilver in much larger proportion than the ordinary sluice without it.

"The saving effected from tailings in one instance is equal to twenty-five per cent., and might be increased; the amount of economy, of course, depending upon the extent to which the under-current is employed."

Hydraulic Mining. § 119. "Hydraulic mining" is that mining where a stream of water, led down from a considerable elevation through a hose, is thrown by the pressure with great force upon the dirt, which is thus loosened, dissolved and washed down into the sluice. The hydraulic power is used to save digging with shovels, to remove the dirt and dissolve it more quickly than could be done in a sluice. "Hydraulic mining," as it is called, is not a process of washing dirt, but of preparing it for washing. The dirt of all hydraulic claims is washed in sluices. The force of the hydraulic stream, sometimes under a pressure of two hundred perpendicular feet of water, is so great that, if it should strike a man, it would kill him instantly; and striking a bank of dirt, it tears it down more rapidly than could two hundred men with picks and shovels. The hydraulic can be used to advantage, only, where there are deep placer diggings, with a channel to lead away the water from the bottom of the claim, and where an abundant supply of water can be obtained at the surface of the claim. In such places, a reservoir for water, sometimes not containing more than a hundred gallons, is placed at the top of the claim, and a hose of heavy cotton or linen duck leads from the reservoir to the bottom, where the water is thrown from the pipe against the earth, near the bed-rock, and that being washed away, the bank above comes tumbling down in great masses, sometimes hundreds of tons at once. The stream of water not only tears down the bank, and carries it to the sluice, but dissolves much of the dirt before it enters the sluice. The amount of dirt that can be washed with a hydraulic depends greatly on circumstances, such as the amount of water, the fall, the character of the dirt, and the season. In winter, more dirt can be washed than in summer; because the earth is then wet through, and dissolves much more readily. Tough clay is very stubborn; sometimes lumps of it, a foot or even two feet in diameter, will roll through the sluice, carrying off much gold and amalgam, which they have rolled over and caught in their course. Large boulders, in some claims, interfere greatly with the progress of washing; these cannot be laid to one side, but must usually be sent down the sluice, either entire or in pieces, after being broken up with hammers. The amount of water used in a hydraulic claim, is from forty to three hundred inches.

Hydraulic miners have had much difficulty with their hose; for the strongest duck and leather would not hold more than about eighty perpendicular feet of water, and would be worn out after a few months of use. The latest mining invention, called the "crinoline hose," is to surround the hose with galvanized iron bands, which are about two inches wide, and from one to three inches apart. These bands are connected together by four ropes, which run longitudinally from one band to another. The crinoline hose will support a head of water more than twice as high as that of a common hose.

A friend of mine who has a hydraulic claim in Placer county, washes about twenty feet, cubic, in twenty-four hours, with a hundred and fifty inches, and six men, running day and night, with three men at work at a time. His dirt is of about medium character, in regard to difficulty of washing, and he has no stones or boulders too large to pass entire through his sluice. Welsh & Co. "piped off," at Indian Hill, Sierra county, a piece of ground forty feet wide by eighty feet long, in six days, using two hundred inches of water, and employing ten men. They paid four dollars per pay to their workmen, making two hundred and forty dollars of wages; three hundred for water, at twenty-five cents per inch; and, perhaps, one hundred dollars more, for waste of quicksilver, wear of sluices, etc., making a total expense of six hundred and forty dollars. They obtained $3,000 in gold dust, leaving a profit of $2,350, for the week's work. The amount of dirt washed having been 224,000 cubic feet, and the gold yield $3,000, we know that there was an average of a cent and a fifth in every cubic foot; and as the claim would have been profitable, if there were much dirt in it, at half the amount which it did yield, we may infer that hydraulic claims, favorably situated, and containing dirt easy to wash, will pay if there be six-tenths of a cent in a cubic foot of dirt. In 1849, miners would not wash dirt with a rocker, unless it contained a bit, (twelve cents) to the pan, which would contain about half a cubic foot of dirt. Hydraulic miners say that dirt containing a "color" to the pan will pay—that is if one speck of gold, no matter how small, can be found, on an average, in every pan, the claim is worth working. Some claims, in which "a color" could be obtained only once in five or six pans, have paid well. The hydraulic miners would always like to work their claims to the bed-rock; for ordinarily the dirt at the bottom is the richest; but sometimes, for want of a channel to carry off their tailings, they cannot get within a hundred or two hundred feet of the bed-rock.

Tunnel Mining. § 120. The tunnel occupies an important place in the placer mining of California. The word, according to the general usage of our language, as explained by Worcester, means a "subterranean passage for a canal or road," running through a piece of ground, from daylight to daylight. But in California usage, a mining tunnel is an adit or entrance nearly horizontal to a mine, or a horizontal drift, carried out from a shaft. Mining tunnels run *into* hills and mountains, not *through* them. They usually have a slight inclination upwards as they enter the hill; for the double purpose of drainage and facilitating the hauling out of the dirt. The length varies from fifty feet to a mile; the cost, from two dollars to forty dollars per foot. In some places, the rock is hard enough to keep in position without support; but usually wooden supporters are used to keep the dirt overhead from falling in, and sometimes the sides must be boarded up. Tunnels are from three to six feet wide, and from five to seven feet high.

Tunnels are always, or with very rare exceptions, made by companies, usually numbering eight or ten men. They claim a large amount of mining ground, and have one entrance to it. Only two or three men can work in digging a tunnel, at a time, to advantage; one to dig dirt; another or two to take it out. For one hundred or one hundred and fifty feet, a wheelbarrow is used to remove the dirt; after that, a wooden tram-way is laid down, and a car is used. Two men may be engaged in hauling out the dirt, each with a car. In such case, there are switches where the empty car can turn out to let the loaded car pass. Work in a tunnel is as pleasant by night as by day; and it is very common to employ relays of men in tunnels. Sometimes there are two sets of hands, to work twelve hours each; sometimes, three sets, eight hours each. Frequently tunnel companies—and the same may be said of nearly all other companies engaged in placer mining—are made up, partly of merchants, physicians or lawyers, and partly of laborers. The latter do the manual labor, and the former supply them with tools, provisions and some cash. The laborers are under the control of a superintendent, who treats all the workmen as if they were hired hands; but he must give employment to members of the company desiring it, in preference to outsiders. On any other system, such companies would soon break up. Sometimes two companies, driving adjacent tunnel claims, unite to dig one tunnel, through which they can both send their dirt out to be washed. Once having reached the pay-dirt, a surveyor is called, and he marks out the line of their claims, and a drift is usually run by the compass along the dividing line. The dirt from this drift is washed, on joint account; and from that time forward, each company knows where its dirt lies; each keeping on its own side of the drift. A joint tunnel for two claims is usually cut along the dividing line. Both companies use the same main tram-way for hauling out their dirt. They may also have a joint sluice, each using it for a couple of days, and cleaning up carefully at the end of that period. A sluice will, in one day, wash all the dirt obtained in six or eight; so that there is little inconvenience about a joint-stock sluice, while much expense in sluice-boxes and water may be saved. In such case, each company has a bin or platform on its side of the mouth of the tunnel, where the pay-dirt is thrown until the proper time for washing. The barren dirt, and stones dug in the tunnel, are often put into the drifts to support them, and protect them from caving in.

After the pay-dirt has been reached, the work is called "drifting," and numerous posts are required to support the roof. Sometimes pillars of the pay-dirt are left.

Many tunnels furnish enough water to wash their dirt.

In places where the tunnel has to be blasted, the expense is from twenty dollars to forty dollars per foot. Two men will blast from five to thirty inches per day.

In earth or soft rock the expense is from two to ten dollars per foot.

Shaft Mining. § 121. Shafts are not much used by placer miners, in California; but they are occasionally necessary to get access to deep claims, where neither the hydraulic process nor the tunnel is applicable. These two have great advantages over the shaft, which is a perpendicular hole, from four to six feet in diameter and fifteen feet deep, or more. The hole is dug by one man, and usually the dirt is hoisted by one or two others, using a windlass and a bucket or tub. When the depth is great, a horse is frequently used for hoisting. The deepest shaft in the State is more than six hundred feet; but there is probably no shaft, now used for placer mining, more than two hundred feet deep. The pay-dirt having been reached, the miner "drifts" sideways from the shaft, and supports the dirt overhead with posts. The pay-dirt is all hoisted up in buckets, and must be extremely rich to pay for this slow work, as well as for the sinking of the shaft.

Shafts are, perhaps, more frequently used in prospecting than in mining. For instance, when it is known that a hill contains in its center a deposit of rich dirt, a shaft is sunk from the top to ascertain the depth of the deposit; so that a tunnel may be commenced at the right level. Ordinarily, the tunnel will have to go farther than the shaft; and sometimes through hard rock; and it will be useless, if too high or too low; whereas, the shaft runs only through dirt, in placer diggings, and is in such cases much cheaper. Sometimes, too, the shaft is dug to ascertain whether there is any gold in a hill or flat, where it is proposed to use the hydraulic process.

The artesian well auger has been used on several occasions for prospecting. It is far cheaper to bore an artesian hole, than to sink a shaft; and in many places it would do as much good in ascertaining the depth at which various strata lie from the surface. There is the objection, however, that the artesian auger makes only a small bore, and brings up little dirt, in which gold might not be discovered, even though there were enough to pay richly. Gold was discovered in the dirt taken from an artesian well, in San Francisco, one hundred feet, and in a well, in Stockton, eighty feet below the surface.

River Mining. § 122. River mining, or mining in river beds, differs from mining in hills, flats and ravines, because of the different nature of the ground. Before the pay-dirt can be reached, the water must be turned from its course and led away in a ditch or flume, or confined to one side of its bed by a wing-dam. Ditches are very rarely used, because of the very steep, rough and rocky character of the river banks, in the mining districts; so, usually, a flume is resorted to. The river is dammed above the place supposed to be rich, and a flume is built to carry the water the length of the claim, at least. A river is seldom flumed for a less distance than three hundred yards. A wooden flume, with bottom and sides of plank, and a regular descent, ten feet wide and three feet deep will carry all the water of a large river, in the dry season. The dam, however, will always leak a little, and so will the flume; and, therefore, the claim will not be dry enough to work without some pumping. This is done with rough board pumps, driven by water wheels, which are placed across the flume, as long as it is wide. These wheels have a shaft running out to the pumps, and driving them. The dirt is washed in sluices, supplied with water from the flume. Large stones and boulders are frequently found in the beds of rivers, as well as on the bars; and these must be removed; for under them the dirt is usually richer than elsewhere. Derricks are used for their removal. The dams are made with brush, stones and dirt. The construction of the dams and flumes is usually commenced about the middle of June, or first of July, when the high water, caused by the rains of the winter and the melting of the snows high up in the Sierra, has gone down for the season. River mining in California closes with the first heavy rain of the rainy season, which carries away dam and flume, and fills the flume with stones and barren dust. This rain may come in the beginning of November; but it is not expected before the middle of December. There are, therefore, nearly six months of the year in which river mining may be done. In many places, a wing-dam which runs half way across the stream, and then down with the middle of the bed, is cheaper and better than a flume.

Mr. Capp, writing in the *Bulletin* of this kind of mining, says:

"These are always joint stock operations, large numbers of

miners engaging in them. The length of time after the melting of the snows in the surrounding mountains, during which the rivers are at a stage low enough to permit their being worked, varies according to the dryness of the season, from six to eight or ten weeks. The expense of grading and building the flume out of lumber, and also for pumping machinery, and the hands required to strip off the light sand, is very heavy. Yet all this has to be done before any of the gold can be reached. A sudden rain storm and freshet, or the setting in of the rainy season earlier than was expected, may bring down a body of water which, in a single hour, may break their dam, carry away their flume, and wash off every vestige of their whole summer's work. Or, if the season is favorable, they may find after long labor that their excavations have been made in the wrong parts of their claim; and just as they reach the lead, the water may be upon them, and the season ended. This is itself a small success, for the next year, without useless work, they may at once commence operations at the proper place. But in many places ill-success of one sort or another has attended them every year; and other claims, though fairly tested, have never paid at all. On the other hand, the yields of such claims have occasionally been enormous.

"Upon a single stream, within a distance of a few miles, during the busy season in the fall, four or five thousand miners may be employed. Canvas towns, with stores, express offices, and drinking and eating saloons, spring up like magic, and are noisy and populous. But the first winter rain destroys the whole busy picture. The flumes, wheels and tools, or as much of them as is possible, are hastily broken up, taken out of the river and piled upon the bank, and shortly afterwards sold at auction, and the proceeds divided. The miners pack up their things and hasten to the gulches, creeks and flats, where they have their winter claims, in which they place their main reliance; and the store-keepers, whisky-sellers, gamblers and expressmen hasten to follow their customers. The size and limits of the claims are, however, recorded and well-known, and at the next season, if the laws have been observed, each man and each company may return and claim and reoccupy its own again. The work upon the rivers is done at a season of the year when many other kinds of mining cannot be carried on for want of water; and those who engage in it are mostly thrown out of employment elsewhere at the time, from that reason. Thus the dryer the season, and the greater the number of miners thrown out of employment in consequence of the want of water, the greater are the chances of employment afforded them upon the rivers, and the more likely are their labors there to be profitable.

"Upon the whole, this kind of mining, though more tempting than many others, has also more uncertainty attached to it, and is regarded as a sort of gambling, sure either to enrich or ruin those who engage in it. It must also be confessed that if the expenses incurred in it are set off against the gold actually obtained, though many large fortunes have been made at it, it will not, upon the whole, appear to have been profitable. The reason of this is, that the heavy expense incurred during any one year has all to be again incurred the next time the operation is attempted. To prevent this heavy annual outlay, it has been proposed, in neighborhoods where these operations have been moderately successful, to build or dig large and permanent canals, so that each season very little expense beyond the construction of the dam at its head, will be necessary. Where this is done, fluming operations may be rendered comparatively certain and profitable speculations, and as the amount of water will be unlimited, and the contents of the river beds are continually enriched by the tailings thrown into them, the washing of the whole may prove to be profitable every year."

Beach Mining. § 123. "Beach mining" is done on the sand beach of the ocean. The shore of the Pacific, from 47°, 30' in California, to 50° in Oregon, is composed of auriferous bluffs, with sand beaches at their base. The bluffs are gradually worn away by the surf, and their gold, which is very fine, is deposited in the sand beach below. This sand is very rich in some places; but the gold is so fine that it can only be caught with the greatest difficulty. The sand is washed in sluices and in toms. In some places, where the beach is low, the sand is covered with water most of the time, and it can only be obtained at very low tide and in still weather. The Marysville *Express* thus described the mode of working practiced by a company of miners at Gold Bluff, in 1859 :

"When the company's watchers discover black spots on the sea-coast, they report to the company, and preparations are immediately made for work. The mules are got ready, with their saddle-bags, and are led to the shore. Men and mules thus harnessed are in waiting for the surf, which at times comes very swiftly, striking the bluffs with great force. When the waves recede, leaving the shore uncovered, men and mules hurry out and gather a load of sand, and hasten back in advance of the breakers, and deposit their loads at the wash house, where it is washed in a long tom. The gold does not come from the sea, but from the bluffs, and is washed back by the surf."

Blasting. § 124. Blasting is not unfrequently used in placer mining, to loosen the dirt which is to be washed or dug. For instance, it has been used in hydraulic claims, to shatter very stiff clay, which, after being shattered, is much more easily washed than otherwise. In such cases, from twenty to one hundred kegs of powder are used at a time. A writer in the San Andreas *Independent* says:

"Caving, or breaking down dirt, is more than two-thirds of the expense of working claims. I am satisfied, from some experiments that I have made on a small scale, that 600 pounds of powder, properly placed and confined, will loosen more dirt than the labor of six men, with the tools now in use, will do in one month; while the expense attending is greatly in favor of the powder. The latter will cost, say $165; the labor and tools of six men will cost at least $500, leaving $335 in favor of the powder, besides costing a great deal less for water to run the same amount of dirt off.

"Miners owning deep claims in localities where the water fails in the dry season can employ their time profitably in running drifts, cutting chambers and preparing the wires properly to loosen up their entire claims, or so much as can be worked while the water lasts. In many localities, the benefits derived from exposing the dirt to the weather will more than repay the expense attending the production of that result. It has been discovered, in many instances, where claims would not pay to sluice, to pay good wages when drifted out and the dirt exposed to the air before washing."

"Instead of piping or picking at a hard bank of earth, for perhaps more than half the time, the miner," says the San Juan *Press*, "can now keep his boxes running full of dirt constantly, at a great saving of time and money and labor. The great desideratum with miners has long been, to obtain an uninterrupted supply of dirt, at as small a cost for labor as possible. The application of blasting appears, so far, to meet this want."

Mining Ditches. § 125. A great portion of the water used in placer mining in California, is brought to the places where used by ditches. The auriferous districts are very dry in the summer and fall, and the supply of water is not sufficient in the winter. It is therefore necessary to bring water in artificial channels, from streams high up in the mountains. These channels are, in some places, ditches cut in the surface of the ground; in others, they are carried as tunnels through mountains; and again, they are wooden aqueducts, called flumes, crossing valleys and ravines several hundred feet above the ground, supported by a high wooden frame work, or suspended upon wires. In one or two places, long iron pipes have been used. The ditches and flumes vary in size; but a flume six feet wide and three deep is considered a large one. They are usually made by companies, organized for that special purpose. The company having been organized and incorporated, the route is surveyed, and the ground required for reservoirs, dams and ditch is staked off. The course is usually through unoccupied land, and the right of way costs nothing.

The water is sold by the water companies to the miners, at a fixed price per "inch." It runs out from the side of the flume, through an opening one or two inches high, the water standing inside six or seven inches high above the opening. In some places, it is understood that there shall be six inches "head," in others seven. An inch of water is as much water as will escape, under that head, through an opening an inch square. The water, however, is not measured with precision. In the morning, there may be fourteen inches of head, and with the additional pressure the amount of water escaping through an orifice an inch square increases greatly; in the evening, there may be a head of not more than three or four inches.

The price of water is usually fifteen or twenty cents per inch per day. In a few places, it is as low as twelve cents; in a few others, as high as forty cents.

CHAPTER IX.

PROCESSES OF QUARTZ MINING.

Comparison of Quartz with Placer Mining.
¿ 126. Quartz mining requires more capital, more expensive machinery, and more scientific knowledge than placer mining. In the latter, mechanical processes have a more prominent place than in the former. The placer dirt is dissolved in water; the quartz rock is pulverized by machinery. A placer claim containing ten cents in a cubic yard of base material may pay; a quartz claim will not pay, in California, unless it contains eight dollars in a cubic yard; and very few veins are worked that do not pay ten dollars. Placer mining is not only much cheaper, because the earthy matter may be so readily dissolved in water, but because the gold is coarse and is more readily caught in sluices. The gold in quartz must be reduced to a very fine powder; otherwise it could not be separated from the rock. It is estimated that from ten to twenty per cent. of the gold is lost, both in placer and quartz mining. In sluices, much of the fine gold is carried away by the water; so too in quartz mills, and in the latter an additional loss is caused by the attrition under the stamps, for whenever a grain of quartz rock is rubbed forcibly against a particle of gold, the latter leaves a portion of its color and substance on the former.

The mode of working will depend upon circumstances, such as the nature of the vein, whether rich or poor, large or small, hard or soft; whether the particles of gold be fine or coarse; whether the vein contain sulphurets or not; whether the vein be easily accessible with wagons; whether wood and water be abundant, and so forth.

The main processes in quartz mining are, quarrying the rock, pulverizing it, and separating the gold from the powder.

Quarrying the Rock. § 127. The quarrying of quartz is done like the quarrying of other rock, and therefore requires no special explanation here. Blasting is ordinarily necessary. The instruments used are the pick, shovel, crowbar, drill, and blasting powder. While quarrying, the miner occasionally assays the rock, by pulverizing some of it and washing it in a horn spoon, as described in § 96. An experienced miner soon learns not only to know whether rock will pay to work, but how much it will pay.

Pulverizing the Quartz. § 128. The pulverizing of the rock is done with a hand-mortar, a large stone, an arastra, a Chilean mill, a stamping mill or horizontal stones.

The hand-mortar is used when the quantity is small, or the rock very rich in coarse gold. In the latter case, the quartz is easily broken up, and most of the gold is picked out with the fingers.

Pulverizing with a Stone. § 129. A large stone is sometimes used by Mexican miners. They fasten a granite boulder, weighing from sixty to one hundred and twenty pounds, to the end of a long, strong pole—usually the trunk of a small tree, fifteen feet long and five inches thick—putting the boulder in a fork at the upper end, and then they lay the pole across a support or fulcrum, so that the stone may be raised and allowed to fall by a man at the other end, acting in a see-saw manner. Under the boulder is a rock on which the quartz is placed to be pulverized.

The Arastra. § 130. The arastra is a mill to pulverize and amalgamate auriferous quartz, and it is considered one of the best instruments for doing that work. It is a circular bed, from eight to twenty feet in diameter, paved with stones, in which quartz, broken into small pieces, is placed and ground to powder by dragging a muller or large stone over it. Arastras are of two kinds, the "rude" and the "improved." In the rude arastra, the bed is paved with unhewn, flat, hard stones, laid usually in common dirt, sometimes in clay or cement. If laid in cement, numerous crevices are left between the stones. In the improved arastra, the paving is done with stones nicely hewn, of greenstone or very hard granite. Around the bed is a wall of stone a foot high, and in the center of the bed is a post to which are fastened two arms, and to these, extending across the bed, are fastened with chains a couple of mullers, which are to be dragged round and round in the bed to grind the quartz. At the end of the arm is hitched a mule, which has a circular path along the edge of the bed of the arastra. Sometimes only one muller is used; sometimes two, and in these cases one mule or horse is sufficient to do the work; sometimes there are four mullers, and then two mules are necessary. The mullers should be long and flat, of very hard material and coarse grain, and weigh from five hundred to eight hundred pounds. They are so hung as to have the forward end about an inch above the bed of the arastra, while the hind end drags.

When the arastra is to be used, a batch of quartz broken into fragments of an inch in diameter or less, and amounting in all to four hundred or five hundred pounds, is put into an arastra ten feet in diameter—the amount of quartz being greater if the arastra be larger. The mule is then hitched up and started, and if the quartz be not very hard, it will be ground to a fine powder in about four hours. Water is then poured in gradually, while the mule still continues his labor, until the pulverized quartz and water have taken the consistence of cream. Quicksilver is now added in the proportion of an ounce and a quarter to every ounce of gold in the quartz. The amount of gold is guessed at from previous examinations and workings of the rock. The grinding continues for an hour longer, and the quicksilver is thereby broken up into very fine particles, and distributed all through the pulp, and brought into contact with every portion of it. The amalgamation having been completed, a large amount of water is let into the arastra and the mule is driven around slowly, to dissolve the thick pulp in the water and allow the particles of quicksilver and amalgam to settle to the bottom. A quarter or half an hour is sufficient for this, and then a gate at one side of the arastra is opened, and the thin mud is carried away by a stream of water running in on one side and out on the other, leaving the gold and amalgam on the bed of the arastra.

In the rude arastra, the quicksilver amalgam settles down into the crevices between the stones of the pavement. After the thin mud has all been swept away, another charge of quartz is put in and the same process is repeated. At the end of the week, the paving is dug up, all the dirt between the stones and attached to them and under them is carefully collected and washed to save the amalgam, which is then retorted.

The arastra is an excellent device for catching the gold in quartz rock by amalgamation. The thick consistence of the pulp, the slowness of the motion of the muller and the complete supervision of the miner over the whole process, are extremely favorable to the main object of amalgamation. It has not unfrequently happened that Mexicans in California have made fifty and sixty dollars per ton from quartz with arastras, and Americans have bought their claims, erected large stamping mills to crush six or eight tons a day, expecting to make a clear profit of several hundred dollars daily; but have found their investment very unprofitable, because the quartz would not yield more than ten or twelve dollars per ton to rapid and " improved " modes of amalgamation.

The arastra, besides its thoroughness, has the advantages of being cheap, simple, easily managed, and better suited than any other mode of amalgamation for a place where water is scarce. Its great disadvantage is that it is so slow. The arastra is sometimes used for amalgamating tailings that have run through other quartz mills.

The Chilean Mill. § 131. The Chilean mill works on the same general principle with the arastra; it pulverizes and amalgamates at the same time, but instead of grinding with a drag-stone, it uses a large stone wheel, which runs round and round in a bed similar to the bed of the arastra. The mode of operation and of amalgamation is very similar to that of the arastra.

An improved Chilean mill is made with two large iron wheels running on one shaft, which turns on a pivot, and the two wheels run round in a little iron basin, into which a small stream of water runs regularly, with an escape on one side for the water and light dirt.

The Square Stamp. § 132. Ninety-five per cent. of the quartz pulverized in California is crushed by stamps. These are of two kinds, the square and the rotary.

The square stamp is a wooden shaft about eight feet long, and from five to eight inches square, shod with an iron shoe weighing from a hundred to a thousand pounds. The stamp stands perpendicularly, and it crushes the rock by falling with all its weight from six to eighteen inches. It is raised by a cam or tooth in a revolving horizontal shaft, and the stamp is lifted up and allowed to fall at every revolution of the shaft. A number of the stamps stand side by side in the stamping mill, and they rise and fall consecutively. The stamps fall into the stamping box or battery. The quartz is broken to the size of a hen's egg or a pigeon's egg before it is thrown into the battery. Quartz is crushed either wet or dry; in the former case a stream of water runs constantly into the battery; in the latter the rock is crushed without water. The battery is surrounded by a screen or sieve of wire-gauze or perforated sheet-iron, which allows the quartz which has been sufficiently pulverized to escape.

The Rotary Stamp. § 133. The rotary stamp has an iron shaft, which receives a revolving motion as it falls. The rotary stamp is preferred to the square, and the latter has been discarded in those mills which have the best reputation for excellence of management. The rotary stamp, according to report, crushes more rock with the same expenditure of power, wears away less of the shoe, and has several advantages in consequence of having more room in the battery.

Horizontal Stones. § 134. Stones, similar in size, shape and mode of operation to the burr mill stones for grinding flour, have been used for grinding soft quartz, and for pulverizing quartz that had been calcined to drive off the sulphur from iron and copper pyrites.

Separation of Gold. § 135. The principles applied in the separation of gold from pulverized quartz are the same with those used in the separation of placer gold from the clay and sand in which it is found. The only difference in the process is caused by the greater fineness of the material upon which the operation is performed. The means of separation are mechanical and chemical; the latter consisting of amalgamation in various shapes; the former, in arresting the gold on a rough surface by the aid of its high specific gravity.

Appliances for Separating. § 136. The principal mechanical appliances for separating gold from pulverized quartz, are the blanket, hide and sluice; the chief chemical appliances are amalgamation in the battery, the amalgamating copper plate, and a multitude of amalgamating pans.

There is no one system of catching gold generally adopted in the quartz mills of the State; indeed, there are scarcely half a dozen that treat their pulverized quartz in the same manner.

The Blanket. § 137. The blanket used in quartz mills, is a common, coarse, gray woolen blanket; the coarser and rougher the better. The blanket is laid down in a trough six feet long and from fourteen to thirty inches wide, and as the quartz is carried over it by the water, the gold is deposited. Some very good quartz miners put the blanket next the battery, and consider it an important point to have the gold pass the blanket before being subjected to amalgamation. Others amalgamate first, and give the blanket a subordinate place. The blanket is taken up and washed in a vat from time to time, according to the amount of gold which it catches. In some mills the blankets are washed twice every hour; in others only twice a day.

The Golden Fleece. § 138. The hide is a cow or calf skin with the hair on, with the hair lying against the current. The purpose of the hide is the same as that of the blanket. Sheep skins with the wool on, reminding us of the Golden Fleece, have been used in a similar manner, but without satisfactory results.

The Sluice. § 139. The sluice used in quartz mills is similar to the placer board sluice. Sometimes it is small and short, with transverse riffle bars; sometimes long and large. Quicksilver is ordinarily used in the quartz mill sluices. In those places where the quartz contains much pyrites, it is common to use the sluices for collecting the sulphurets of iron and copper, which are saved to be operated upon at some future time. This operation is styled " concentrating the tailings," and the material obtained is called " concentrated tailings."

Amalgamation. § 140. Amalgamation in quartz mills is effected by very many different processes.

In the Battery. § 141. The first method is to amalgamate in the battery. Two ounces of quicksilver are put in to catch one ounce of gold, and the amalgamation proceeds more rapidly and more gold is caught when that proportion is preserved, than if there be too much or too little quicksilver. In the Sierra Buttes mills two-thirds of the gold is caught in the batteries.

Copper Plate. § 142. The amalgamating copper plate is very extensively used. It is a copper trough, from three to ten feet long; sometimes nearly smooth on the bottom, sometimes provided with riffles and basins. The surface of this trough is covered with quicksilver, which amalgamates with copper as with gold and silver. The copper keeps a surface of quicksilver thus constantly ready to catch the gold. A good description of an amalgamating copper plate may be found in § 113.

Amalgamating Basins. § 143. Amalgamating pans or basins are made in many styles. One is like an arastra; another is like a Chilean mill; a third is a large bowl with a stick that stirs round and round; a fourth is a large bowl with square compartments in the bottom, every compartment containing quicksilver, and the bowl is shaken violently so that the pulverized quartz is thrown successively from one compartment to another; a fifth is a large bowl, revolving, with a slight inclination, on an eccentric axis; and in a sixth the pulverized quartz is forced through mercury. The amalgamating basins always contain a supply of loose quicksilver. The arastra and Chilean mill are used sometimes to grind the powder finer, as well as to amalgamate. While amalgamating, the miner occasionally washes some of the pulverized rock in a horn spoon, to judge of its richness and to know how much quicksilver must be used.

General Remarks. § 144. The devices which I have here mentioned for separating the gold from pulverized quartz, are but a few of a great multitude, most of which do not require a particular description here. Mr. Capp wrote thus, in 1857, about the methods then used in the quartz mills of Grass Valley:

"Some persons prefer to separate as much of the gold as possible without mercury. Others believe the sooner the gold is brought in contact with mercury the more certainly is it secured and prevented from passing off. Accordingly, in one instance, in Wiggain's mill, at Nevada, I found that mercury was put into the batteries along with the quartz, thus catching a large part of the gold there. This plan, however, has generally been abandoned in this vicinity. In the Empire mill, at Grass Valley, there was placed a pan of mercury directly under the batteries, so that all the quartz, as it escaped through the sieve, first fell from a spout into this pan, and left a large portion of its gold there. In the French mill, before going any further, after passing the sieve, an opportunity is given to the gold and heaviest portion of the quartz to settle in a box, from whence it is taken along with the blanket washings to the amalgamator and thence to the grinder, which is used in place of the Chile mill. But in most of the mills, after passing the sieve, the quartz is carried by the water which splashed it out of the battery down a trough ten or fifteen feet in length, with a fall of eight inches or a foot. A coarse, long-haired blanket is placed in the bottom of this trough, and affords just sufficient obstruction to the water to cause a large portion of the gold to settle upon it, while ninety-five per cent. of the rock is carried on beyond it. At the Mount Hope mill, the blanketed trough is divided into two lengths, and between them a pan of mercury is placed, so that a portion of the gold is caught there, and what passes over the second blankets has thus been in contact with mercury. The rule generally is, not to permit mercury to touch the gold before it passes the blankets. This is because in coating the particles of gold the mercury renders them globular, destroying their natural angles, thus rendering them less likely to be caught by the hairs of the blanket, and more liable to be rolled and carried off by the water.

"Below the blankets a variety of amalgamators and other contrivances are used, to cause the gold to settle and bring it into contact with mercury. At the Gold Hill mill, below the blankets small falls are arranged, so that the current is broken and the gold assisted in finding its way through the quartz to the bottom. Afterwards it is received into tanks about eight feet long, where nearly all the current is stopped, and the heaviest portion settles at the upper end. That quartz which has settled within a foot or so of the upper end is called a 'beading,' and is taken out with that from the falls above the blanket washings to be amalgamated and then ground in the Chile mill. In the Empire mill, the quartz, after passing over the blankets, falls into a pan of mercury, and from that into another, and so on through five pans. Thence it passes to tanks similar to those mentioned above. The muddy water which passes over, and the rest of the quartz which is washed out when the headings have been reproved, go through a Cram's riffle, which is merely a box with compartments arranged as a series of falls of about six inches each, under which mercury stands to the depth of an inch. A partition in each compartment extends across and to within a short distance of the bottom, so that the water and quartz must fall behind it, and in order to escape have to pass under the lower edge close to the mercury, and then rises in front and falls into the next riffle. After thus passing four or five riffles, it is allowed to escape to the pile of tailings or down the creek, but passes, in doing so, through a trough which has quicksilver in it. In other mills, the quartz is received from the blankets into a Stetson's Amalgamator, which is a chest of shallow drawers, each of which is perforated with half-inch holes, arranged so that those in one drawer are over the centers between the holes in the one below. Around these holes are small grooves, which are filled with mercury, into which the gold and quartz must fall, and the former is caught and retained while the latter passes on, and is carried by a trough to the tailings pile.

"The ground quartz and gold which has settled on the blankets and in any riffles or other contrivance above where mercury is used, is removed and worked with water by a man using his fingers, or as at the Gold Hill mill, with a wheel having its circumference covered with projecting claws, passing between similar ones arranged in the trough or basin, which is partially filled with mercury. This causes the largest portion of gold to be caught and covered by the mercury into which it sinks. Thence the water carries the quartz along a trough in which are riffles or cuts filled with mercury. The trough leads it into the basin of the Chile mill, where it is ground as fine as flour in connection with mercury. The grains of quartz and pyrites of iron being thus disintegrated, any particles of gold which have been held in them are liberated, and their surfaces cleaned and polished. The mercury is then able to attack them, and an amalgam is formed. The quartz and pyrites thus finely pulverized gradually mix with the water and pass off as a muddy stream. Sometimes amalgamators are placed below the Chile mill, through which this muddy water has to pass, so that another and final opportunity is afforded for any free gold that may have escaped to unite with the mercury. But very little is ever collected in amalgamators below these mills, and little attention is paid to what escapes them. The Chile mill consists of two heavy iron wheels, with broad flat tires, rolling slowly in a cast iron basin. At the Gold Hill mill, stirrers run in front of the wheels, continually throwing the pulp directly in their path.

"At the Empire mills the blanket washings are run a second time over blankets, so as to let a portion of the less valuable and coarser parts escape into the ordinary amalgamation. The residue which is retained is then amalgamated in a trough as usual, but nitric acid is used in connection with the mercury. This, it is believed, cleans a portion of the gold, on which there is often an insoluble coating which prevents the mercury from attacking it, and also by dissolving some of the sulphurets frees the gold which they contain, and renders its amalgamation and separation more perfect. The additional expense is very trifling, and the proprietors are satisfied with the results.

"The action of mercury upon gold is not always instantaneous, so that when the two are in contact for any length of time, much gold is apt to be collected which remains entirely free and passes off with the water and fine quartz when a continual motion is kept up, as in a trough or any kind of amalgamator in which a constant current is kept up. Grinding and pressure also greatly facilitate the action of the mercury. In the Brazilian quartz mines this is well understood, and the blanket washings—the richest portions of the pulverized quartz—are kept in contact with mercury for twenty-four hours, during which time a constant grinding is kept up. For this purpose the fine quartz is placed with water and mercury in barrels, which turn continually upon a horizontal axis. The result is said to be very satisfactory; but the operation requires more time and power than many think would be profitable in California. I am convinced, however, that some such means, affording a more lengthened contact with the mercury, will be resorted to hereafter in place of any of the many instantaneous amalgamators now in use, and that much of the gold which now escapes will in this way be secured.

"In the French mill, as in other mills, in front of the stampers a perpendicular sieve is arranged, and the water which is supplied in front of the stamps splashes when they fall, carrying with it the pulverized quartz. That portion which is fine enough passes through the sieve with the water. In most mills the pulverized quartz passes freely away from the stampers into the troughs, but here an opportunity is given for the heaviest portion of it to settle in the boxes immediately in front of the sieves. The greater part, however, is carried by the water through an aperture left for the purpose, and then passes, as usual, through a trough about fifteen feet in length, lined with blankets, on which a portion of the gold settles, and having a fall of eight or ten inches. At the end of this trough the stream is divided, and passes into a set of three small bowls, about eight inches in diameter, and containing mercury. Small arms, moved steadily by machinery, continually revolve in each of these bowls, and bring another portion of the gold in contact with the mercury. The bowls are arranged one below the other, so that the quartz and water pass through all three of them. Leaving the bowls, the quartz is then carried by the water through a Stetson's amalgamator of six drawers, where still another but smaller portion of the gold is saved. Finally, it is carried off to the pile of tailings, but the troughs in which it moves are also quicksilvered and retain a little more of the gold that is left. Thus the bulk of the rock that is crushed is disposed of, but by far the greater part of the gold is mixed with that which remained in the trough in front of the stampers, or has lodged on the blankets. Every twenty or thirty minutes the blankets are changed and rinsed out in water, that their surfaces may remain rough and hairy and not become clogged by the fine quartz, and so fail to afford a lodgment for the particles of metal. Whenever the blankets are changed, the quartz in front of the stampers is removed and thrown with that washed from the blankets. Then, in small quantities at a time, it is placed in a trough upon which a stream of water pours, regulated by an ingenious mechanical arrangement so as to carry the quartz always at the same rate through a series of stirring bowls containing mercury, and similar to those before described. Thence the quartz is taken to the mill in which it is ground, to crush the coarser particles and the iron pyrites, which may protect a portion of the gold from contact with the mercury. Thence it passes, as did the other, through Stetson's amalgamators and quicksilver troughs, which carry it also to the pile of tailings.

" In other establishments, the mill used in grinding the pyrites and coarser parts of the blanket washings is the ordinary Chile mill; but here a new apparatus is employed that has lately been invented and patented by Mons. Chavanne, the manager. For this he claims many advantages; the principal one, however, is that it operates more rapidly, and will grind a larger quantity of tailings per day than the Chile mill, also doing the work quite as perfectly. It will be seen that the two processes of separation are used, by which the quartz as pulverized is divided into two portions, viz.: that which remains in the trough in front of the stampers and on the blankets, and that which passes on beyond them and goes directly through the series of amalgamation first described. To exhibit the effects of the different parts of the process, M. Chavanne very kindly furnished me with the following figures:

"Of a ton of quartz that is crushed, ten per cent. settles in the trough, and five per cent. in the blankets. The remainder passes off with the water through the amalgamators. Suppose that a ton of quartz yielded forty dollars, by the use of the apparatus above described; the gold would then be collected as follows: In the quartz which settled on the trough in front of the stampers and upon the blankets, would be thirty dollars, or seventy-five per cent. of the whole; in the stirring bowls, just below, in combination with the mercury, three dollars, or seven and one-half per cent.; in the Stetson's amalgamator, below them, one dollar and seventy-five cents, or four and three-eighths per cent.; and in the quicksilver trough beyond, seventy-five cents, or one and seven-eighths per cent. Then the thirty dollars has to be separated from the quartz with which it is mixed, and, in accomplishing this, the machinery used acts as follows: In the three bowls would be obtained twenty-five dollars, or eighty-three and one-third per cent. of this portion, or sixty-two and a half per cent. of the whole yield. Of this, three-fourths would be in the upper bowl. In the mill below, four dollars, or thirteen and one-third per cent. of this portion, or ten per cent. of the whole. Perhaps the remaining fifty cents, or one and a quarter per cent., would be found in the amalgamator and trough below.

"This process is considered one of the most perfect in use, on account of the many ways adopted to bring the mercury in contact with the gold, all of which, it will be seen, is amalgamated. It is claimed that by using this process, rock which in 1856 only yielded thirty-four or thirty-five dollars per ton, now yields at least forty dollars. The stirring bowls are thought by the proprietors to be better than the ordinary riffled trough amalgamator, because there the tailings are stirred into the mercury by a man, with his fingers, and though he may operate very regularly for an hour or so, he is liable to become tired or careless. The regular motion of machinery avoids this danger."

Sulphurets and Amalgamation. § 145. The sulphurets of iron, copper and lead are frequently found in auriferous quartz, in considerable quantities, and by their presence they prevent amalgamation, and thus cause the escape of a large portion of the fine gold. The chemical causes of the refusal of the quicksilver to catch gold when sulphur is present, are not fully understood; but the fact is undeniable. Auriferous quartz that contains three hundred and four hundred dollars to the ton, will, if it contain much sulphurets, rarely yield more than twenty or thirty dollars, in very fine gold, to the amalgamating process. The treatment of the sulphurets becomes, therefore, a matter of very great importance to the quartz miner. The sulphurets are decomposed by a high heat, and by long continued exposure to the atmosphere; and when decomposed, they no longer present any obstacle to amalgamation. But the roasting of the quartz, and the exposure of tailings to the atmosphere for a long time, are both expensive, and in most cases will not pay. At many mills, the tailings containing sulphurets are all saved in heaps, for some future time, when they may be worked with a profit.

Quartz Mining as a Business. § 146. Quartz, like all other branches of mining, is a very uncertain business, paying immense profits to some, and causing great losses to others. It is that kind of a business in which men, as a general rule, should not invest money unless they have made the working of quartz their study, and can themselves be present to see how the work is done. But to industrious men, who superintend their own claims, and have much business talent and a thorough knowledge of quartz mining, there are few occupations that promise greater profits, than the crushing and amalgamation of auriferous quartz.

The owners of quartz leads have very often erred, in California, in supposing that because they found very rich rock at the surface of the vein, they would continue to find the same proportion of gold throughout the vein; and acting on this presumption have erected expensive mills, which had scarcely commenced operations before the rich quartz all gave out, leaving stuff that would not pay.

For men having little knowledge of quartz mining, and becoming the owners of a rich quartz lead, the safer plan is either to take the rock to a custom mill—that is, a quartz mill which crushes quartz at a certain price per ton, for all applicants— or to use the arastra. This course may be used until enough money is made to build a stamping mill, and until some experience is gained.

The quartz now worked in California pays, on an average, eighteen or twenty dollars per ton; and costs twelve or fifteen dollars to work it. In some mills, the expense is not more than six dollars per ton. The poorest quartz that is worked yields ten dollars per ton, and the richest vein probably does not average over one hundred dollars.

CHAPTER X.

PROCESSES OF SILVER MINING.

Comparison of Gold and Silver Mining. § 147. Silver mining differs greatly from gold mining. The latter metal is found in mechanical combination with baser substances; the former in chemical union with them. Gold is not found in ore; silver is. Gold is separated from the quartz and diluvium in which it is found chiefly, by mechanical or very simple chemical means, and by processes easily learned; silver is found in many different kinds of ore, each of which must be reduced chemically, by intricate processes, varying greatly according to the character of the minerals and the circumstances in which they are found.

Silver Ores of the Pacific Coast. § 148. The silver ores of Washoe, Esmeralda and Coso are nearly all sulphuret of silver, with particles of native silver and gold interspersed. The ores are peculiar, and different from ores of other argentiferous countries, because they bear so great a resemblance to each other; whereas, in Mexico and Chile, every mine has a different kind of ore, and an experienced miner will tell by a glance at a specimen the name of the mine whence it came. South of Coso, the silver ores vary more, and are mixed with lead, copper, antimony and arsenic, in greater proportions.

The Reduction of Silver Ores. § 149. Silver ores are pulverized in the same manner as gold quartz; but beyond that pulverizing, the methods of treatment differ almost totally. The methods applied to silver are scientific and too complex to be described fully in a book so brief as this must be. Besides, it is asserted that several new modes of reducing silver ore have been invented lately in California, for each of which the merit is claimed of great superiority over all methods previously in use. These processes are named after Mr. Bagley, Dr. Veatch and others, their respective owners or inventors, and they are all secret. Their proprietors are well known and held in high respect in San Francisco, and their claims have all found credence among silver miners. The Ophir Company has adopted the Bagley process, and the Central Company is using the Veatch process. The Ophir Company has paid $10,000 cash for the privilege of using the Bagley process, and are to pay two and one-half per cent. of all the silver obtained by it for three years, and at the end of that time they are to have the privilege of purchasing the right of using the process for their claim on the payment of a very large sum of money. The owners of the process, on the other hand, give a heavy bond that the reduction of the ore shall not cost more than thirty dollars per ton—which is very cheap. The inventor of the process says that after pulverization, the reduction will not cost fifteen dollars per ton. Under these circumstances, and for the reason, also, that neither the new nor the old processes have yet been fully tried in Washoe, and adapted to the circumstances of that place, I shall devote but little space to the description of the processes of reducing silver ores.

Silver ore is pulverized with stamps or with an arastra. The arastra is never employed for amalgamating, as in gold mining, but sometimes it is used in washing out the paste after amalgamation.

Silver ores are reduced by many processes, the principal of which are smelting, eliquation, barrel amalgamation, *patio* amalgamation, and the salt-solution process.

In the first two modes, the silver is separated from the other minerals by the assistance of lead; and the metallic mixture of lead and silver thus obtained is purified by keeping it melted at a high heat for some hours, whereby the base metal is oxydized and driven off, leaving the precious metal pure.

In the three last processes, the silver is converted into a chloride, in which condition it is readily separated from baser material. Silver has a very strong affinity for chlorine, and whenever brought into contact with common salt, takes possession of the chlorine contained in it.

The amalgamating processes are based upon the facts that silver, when in a pure state, is caught and saved by quicksilver; that while quicksilver has no influence on sulphuret of silver, which is found in nearly all silver ores, the sulphuret is decomposed by the presence of common salt, the chlorine of which unites with the silver, while its soda unites with the sulphur; that the chloride of silver thus formed is again decomposed by the presence of either quicksilver or iron, both of which metals are preferred by the chlorine to the silver, and the latter metal is left free to be taken up by quicksilver, some of which remains pure, though a portion of it may have been converted into a chloride of mercury.

These are the general ideas of these processes; now for a few details.

Silver Smelting. ¿ 150. In smelting, the ore is mixed with fifty per cent. of lead, or lead ore containing that amount of lead, and ten per cent. of iron, and the whole mass is melted; when the silver is allowed to run out through a hole in the bottom of the furnace. The silver thus obtained is not pure; but it is put into a furnace and kept at boiling heat for some hours, whereby all the base metals are driven off. This process was used in San Francisco during the winter of 1859–'60, to reduce the first lot of ore brought down from the Ophir claim. The company paid $412 per ton for reducing forty tons; but this price included the cost of erecting furnaces, etc. The cost to the smelter per ton, after the erection of the furnaces, was about $125—the main expense being in the lead, which had to be purchased in the metallic form at seven cents per pound, no lead ore being obtainable.

Salt Solution Process. ¿ 151. The salt solution process is as follows: The ore is roasted with five per cent. of common salt, for several hours. When all the silver has been converted into a chloride, the ore is thrown, at a red heat, into a boiling saturated solution of common salt, which dissolves the chloride of silver. The solution is then filtered while still boiling hot—that high heat is necessary to prevent the liquid from depositing the silver upon the earthy matter of the ore—and the metal is precipitated by the addition of a little muriatic acid and some pieces of copper.

Barrel Amalgamation. § 152. In the barrel, or European amalgamation—it is also called the Freyberg process, from the place where it was invented—the ore containing about twenty-five per cent. of iron pyrites is roasted, in lots of five hundred pounds each, in furnaces made for that special purpose. If the ore, as taken from the mine, does not contain twenty-five per cent. of iron pyrites, that mineral is obtained elsewhere and added; if the ore contains more than that proportion, some of the pyrites is picked out by hand, or some of the pyrites is reduced by heat before the commencement of the roasting. The ore is first dried for twenty minutes; then the fire is made hot, and in two hours the mass becomes red hot. The fire is allowed to go down, and the mass is kept at a low heat until the ore is dark in color, and has ceased to emit any sulphurous smell. The heat is then raised again for three-quarters of an hour.

The ore is now taken out and pulverized, and then put through the barrel amalgamation. The barrel is thirty-four inches high, and thirty-four inches in diameter at the head. It stands upright and revolves on a perpendicular axis. About half a ton of the pulverized ore, three hundred pounds of water, and one hundred pounds of wrought iron in fragments an inch square and half an inch thick, are put into the barrel, which is then set to revolving rapidly. At the end of two hours, the barrel is stopped and the "paste" examined to see whether it is of the proper thickness, which should be about that of thick cream. If too thin, more powdered ore is added; if too thick, more water; and the barrel is set to revolving again until it has the proper consistence. Then five hundred pounds of quicksilver are poured into each cask, which must revolve four hours, when it is examined, and if the consistence be not right, water or ore is added; if right, it goes on revolving four hours more, when another examination is made, and then eight hours of constant revolution finishes the amalgamation. The barrel is filled with water, made to revolve slowly to allow the amalgam to settle to the bottom, on the same principle as in the arastra described in § 130, and at the end of two hours the mercury is allowed to run into a pan through a little hole at the bottom of the barrel. The whole mass of earthy matter is afterwards led through a sluice, to catch such amalgam as was not caught in the pan. There are ordinarily a number of these barrels side by side. They will amalgamate two-thirds of a ton each in twenty-four hours. The amalgamation of each ton costs three pounds of wrought iron and half a pound of mercury.

The amalgam is retorted to separate the quicksilver from it. The metal thus obtained usually contains a considerable proportion of copper, antimony and lead. These baser metals are removed by melting the lead in a furnace, keeping it melted six or eight hours, and throwing powdered charcoal on its surface, and the impurities which form on the surface of the silver by the combustion of the baser metals are skimmed off, leaving the silver pure.

The Patio Process. ₴ 153. The patio process, or Mexican mode of amalgamation, is managed in a *patio* or amalgamating yard, closely paved with granite. Many tons may be operated upon at once—from fifty to a hundred tons. The ore is pulverized finely and spread out in the *patio* a foot deep. Over the ore is scattered eight per cent., by weight, of common salt, and the mass is thoroughly mixed together by the tramping of horses. After lying thus a day it is again trodden over by horses for an hour or two ; one per cent. of roasted copper pyrites is added, and it is again trodden. A man now scatters over the mass three-quarters of a pound of quicksilver for every hundred pounds of the ore. The scattering is done by means of a canvas bag into which the quicksilver is put, and by shaking, it is distributed in very small particles. The horses are again driven through the mass for an hour or two. This is done every day until it is found, on examination, that all the quicksilver has been taken up. Then another lot of mercury is scattered over the mass, one-quarter of a pound to every hundred pounds of the mass, when it is again trodden a little every day, until all that quicksilver is taken up by amalgamation, when another lot of mercury—three-eighths of a pound to a hundred pounds of ore—is added, and lying some days, with occasional treading, the amalgamation is complete. During the process, the miner occasionally takes out some of the mixture and washes it to see the condition of the amalgam. If it be grayish white in color, and can be readily moulded with the fingers, it is in a favorable condition. If the mercury be divided and of a dark color with occasional brown spots, there is too much copper pyrites, and lime is added to neutralize it. If the mercury remains fluid and there is little amalgam, there is too little pyrites and more is added. The amount of mercury to be used depends upon the richness of the ore. A pound and three-eighths of quicksilver may be proper for ores that pay thirty-five ounces of silver to the ton. The progress of the amalgamation depends to a great extent on the heat of the weather; if very cold, it will go slowly or not at all. The *patio* process will, therefore, not be used in Washoe in the winter, and probably not in the summer.

After the amalgamation is complete, the mixture is put into a large tub in which there is a perpendicular revolving shaft with arms. A stream of water runs into the tub on one side and out on the other, carrying away the earthy particles and leaving the amalgam to settle at the bottom. The amalgam is taken out, washed again and then retorted.

The Eliquation Process. § 154. The eliquation process is suitable for argentiferous copper ores. The ore is pulverized, and if it does not contain lead, lead ore also pulverized is thoroughly mixed with it, and then it is subjected to a heat great enough to melt the lead, but not to melt the copper. The melted lead carries off the silver with it. A peculiar furnace is used, and no flame is admitted to the ore, for that would oxydize the lead and prevent it from carrying off the silver. Lead melts at 612°, silver at 1873° and copper at 1996° Fahrenheit. It is not necessary to raise the heat to such a degree as would melt the silver alone, for when silver is mixed with lead, even in ore, it melts at a much lower figure than when pure. In the eliquation process, therefore, a heat of about 1000° is sufficient. The process is one of the most expensive and unsatisfactory of all those used in reducing silver ore, but yet for some kinds of ore it is the best known mode of treatment. There ought to be about four times as much lead as copper in the ore, and the operation is most complete when there are fifteen pounds of mixed lead and copper to an ounce of silver. If there be more than twenty ounces of silver in a hundred pounds of ore, much of the silver will be left behind, and the remainder must be put through the process a second time. At least one-tenth of the silver in the ore is left behind after every eliquation.

New Processes. § 155. Several new processes for reducing silver ores, not yet described in the standard books of metallurgy, have been discovered within the last ten years. For notes of these I am indebted to Mr. C. Wennerhold. The mercurial solution process discovered by Von Fuchs, a German chemist, is as follows :

To one hundred parts of finely pulverized ore, add water sufficient to make a thin pulp, which is to be heated in a shallow iron pan. Dissolve four parts of sulphate of mercury in fifteen parts of water, and put six parts of common salt in twelve parts of water; mix the two solutions, and pour them upon the heated pulp. The mixture must be stirred and kept at a boiling heat for a few hours, till a piece of bright iron, when dipped into it for a few minutes, is not covered with metallic mercury. If the iron should be covered with the mercury, the boiling must be continued. When the mixture has been boiled sufficiently, take from the fire and pour in ten parts of metallic mercury. If the ore contains much iron pyrites, it should be roasted before going into the kettle. Before the mercury has been put in, add seven parts of iron filings.

The theory of this process is, that the solution of mercury comes into contact with every particle of silver, and while the sulphuric acid attacks the iron, the mercury takes up the silver. If experiments were to be made with this process in California, where sulphate of mercury is not easily to be obtained, corrosive sublimate or bi-chloride of mercury might be used instead.

The Muldener process, so called because it is used at Mulden in Hanover, is applied for reducing argentiferous copper ore, containing sixty per cent. of copper, twelve per cent. of lead and from one-third to one-half of one per cent. of silver. After being pulverized, the ore is roasted in a double reverberatory furnace, first in the upper, then in the lower hearth. About a pound of coal is consumed to a pound of ore in the furnace. The roasted ore is ground, and then, after mixing eight per cent. of common salt, is roasted again, whereby the silver is converted into a chloride. The ore is now placed in a wooden tank with a false bottom, which acts as a filter, and a heavy column of water resting upon the ore, and gradually filtering through it, dissolves and carries away all the sulphate of soda, made by the union of the sulphur of the ore with the soda of the salt and oxygen of the air. The ore, freed from the sulphate of soda, is treated again in a similar method with a solution of common salt, to dissolve the chloride of silver, which metal is then obtained from the liquid by precipitation with metallic copper. The copper remaining in the ore is smelted out.

The Blaschka process, (so named from its discoverer, W. Blaschka) is a combination of the process of Ziervogel, for the extraction of silver, with that of Plattner, for the extraction of gold.

The main idea of the Ziervogel process, is to form sulphate of silver, and then separate it with water. In this process, the nature of the ore must be very accurately known. If the metallic substance be only sulphurets of silver and copper, a proper roasting suffices. If there be much metallic silver in the ore, the former may be converted into a sulphate with sulphuric acid, with or without roasting, or by passing the fumes of sulphuric acid over the ore while in the reverberatory furnace.

If a sulphuret of arsenic or of antimony be in the ore, the latter must be mixed with an equal amount of copperas before roasting, to prevent the formation of the arseniate of silver. The roasting must be continued until a sample of the ore, taken from the furnace and submitted to the proper tests, shows no protoxide of iron. The ore is then taken out of the furnace immediately; and the sulphate of silver is extracted in large wooden tanks, with water. One part of sulphate of silver requires eighty-eight parts of water to dissolve it. The silver is then precipitated as a chloride with salt, or as a metal with copper.

After the silver has thus been extracted, the ore goes through the very ingenious process of Plattner for separating the gold, by the formation of chloride of that metal. This method, so far as known, has not been tried on a large scale. Great care must be taken in this process, that the ore contanis neither metallic iron nor undecomposed sulphurets, arsenious acid, antimony, sulphate of iron, nor any metallic oxides soluble in water.

After all the silver has been extracted, the ore, mixed with water, is placed on a layer of quartz sand on the false bottom of a large wooden tank, which is closed very tightly, and chlorine is introduced until all the gold is formed into a chloride, which is dissolved in water, filtered, and then precipitated with copperas.

CHAPTER XI.

THE LAWS OF MINING IN CALIFORNIA.

Public Mining Land open to all. § 156. The miners occupy the public mineral lands of California with the express consent of the State, and with the implied consent of the Federal Congress; and they will undoubtedly be protected in all their possessions held in accordance with the State laws and the local mining regulations. The Federal Government owns nearly all the mineral land, and has shown its intention to leave the miners in possession of their claims, by taking no measures to disturb them, by directing the Federal Surveyor to stop his surveys on reaching the bounds of the mineral districts, and by refusing to sell the land, or open it to preëmption for agricultural purposes.

Mineral Land owned in fee. § 157. A small portion of the mineral land belongs in fee simple to private individuals holding under grant from Mexico. Under the mining laws of most mining countries, miners have the right of entering upon private land sand taking out the minerals; but they have no such right in California. The Supreme Court of the State have decided that the ownership of the land carries the minerals with it. The opinion in the case of *Fremont vs. Flower*, says:

"The construction given by the United States to their patents ever since the organization of the government, has uniformly been to the same effect. In several of the States, particularly those carved out of territories ceded by Virginia, North Carolina and Georgia, and out of the territory acquired by the treaty with France in 1803, and by the treaty with Spain in 1819, the title to a large portion of the lands is held under patents from the United States. Some of these patents were issued upon a sale of lands—some of them upon a donation of lands, and some of them upon a confirmation by Board of Commissioners of previously existing grants of the former governments. Patents upon such confirmation were issued to extensive tracts in the territories of Louisiana, Mississippi and Florida, and in the lands, which the patents conveyed, minerals of gold and silver, and of other metals, in many cases existed. Yet in no instance, whether the patents were issued upon a sale, or donation of lands, or upon a confirmation of a previously existing grant, have the United States asserted any right to the mines as being reserved from the operation of the patents. They have uniformly regarded the patents as transferring all interests which they could possess in the soil, and everything imbedded in or connected therewith. Whenever they have claimed mines, it has been as part of the lands in which they were contained, and whenever they have reserved the minerals from sale or other disposition, it has only been by reserving the lands themselves. It has never been the policy of the United States to possess interests in land in connection with individuals."

In 1853, our Supreme Court, then composed of other Justices, rendered a decision that all the valuable minerals in the earth belong to the State, by virtue of her sovereignty; as under the English law, the monarch is owner of all mines as an attribute of his sovereignty. In regard to this point the Court, in their latest decision, say:

" Sovereignty is a term used to express the supreme political authority of an independent State or nation. Whatever rights are essential to the existence of this authority, are rights of sovereignty. Thus, the right to declare war, to make treaties of peace, to levy taxes, to take private property for public uses, termed the right of eminent domain, are all rights of sovereignty, for they are rights essential to the existence of supreme political authority. In this country this authority is vested in the people, and is exercised through the joint action of their Federal and State Governments. To the Federal Government is delegated the exercise of certain rights or power of sovereignty; and with respect to sovereignty, rights and powers are synonymous terms; and the exercise of all other rights of sovereignty, except as expressly prohibited, is reserved to the people of the respective States, or vested by them in their local governments. When we say, therefore, that a State of the Union is sovereign, we only mean that she possesses supreme political authority, except as to those matters over which such authority is delegated to the Federal Government, or prohibited to the States; in other words, that she possesses all the rights and powers essential to the existence of an independent political organization, except as they are withdrawn by the provisions of the Constitution of the United States. To the existence of this political authority of the State—this qualified sovereignty, or to any part of it—the ownership of the minerals of gold and silver found within her limits is in no way essential. The minerals do not differ from the great mass of property, the ownership of which may be in the United States or in individuals, without affecting in any respect the jurisdiction of the State. They may be acquired by the State, as any other proprety may be, but when thus acquired, she will hold them in the same manner that individual proprietors hold their property, and by the same right; by the right of ownership, and not by any right of sovereignty."

The Federal Judges in California seem to hold the same views of the law. In the New Almaden quicksilver mine case, the U. S. District Court, in their decree confirming the claim, say:

"And it is likewise ordered, adjudged and decreed, that the claim and title of the petitioner, Andres Castillero, to the mine known by the name of New Almaden, in Santa Clara county, Northern District of the State of California, is a good and valid claim and title, and that the said Andres Castillero and his assigns are the owners thereof, and of all the ores and minerals of whatsoever description therein, in fee simple: And it is further adjudged and decreed, that the said mine is a piece of land embracing a superficial area, measured on a horizontal plane, equivalent to seven pertenencias, [about fifty acres] each pertenencia being a solid of a rectangular base two hundred Castilian varas long. of the width established by the Ordenanzas de Mineria of 1783, and in depth extending from and including the surface, down to the center of the earth; said pertenencias to be located in such manner as the said Andres Castillero or his assigns may select, subject to the following conditions: first, that the said pertenencias shall be contiguous, that is to say, in one body; and secondly, that within them shall be included the original mouth of the said mine known as 'New Almaden.'"

This decree is made under a law to confirm valid land titles granted by Mexico, but Mexico never gave a fee simple title to minerals; mines were always and still are held in Mexico only while they are worked, and neglect to work them causes a forfeiture and justifies any new claimant in taking possession. The Mexican system is unsuited to the American modes of doing business, and at variance with the principles of American jurisprudence.

The principal tracts of private land containing mines, are the ranchos of J. C. Fremont, in Mariposa county, of the heirs of J. L. Folsom, in Sacramento county, of John Bidwell, in Butte county, of P. B. Reading, in Shasta county, of Charles Fossatt, and the New Almaden Mining Company, in Santa Clara county.

Foreign Miners. § 158. All citizens have a right to mine on any of the public land in the State, without paying any tax or license for that privilege. Aliens are required by law to pay a license of four dollars per month for the privilege of mining. Any person or company hiring aliens to work a mining claim, or renting a claim to them, or interested with them as partners in mining, is responsible for the license money. Private parties, however, have no right to eject aliens from their claims, because of their neglect or refusal to pay the license. The collector of the foreign miners' tax has broad powers to enforce the payment of the tax, including authority to sieze the alien's claim and tools, and sell them at a very brief notice. This law is intended chiefly to levy a tax on Chinamen, and in many districts no attempt is made to collect anything from Irish, German, French and other European miners.

It is very clear to my mind that the whole foreign miner act is at variance with a clause of the State Constitution, (Sec. 17, Art. 1) which says: "Foreigners who are or may hereafter become *bona fide* residents of the State, shall enjoy the same rights in regard to the possession, enjoyment and inheritance of property as native-born citizens." It is not to be denied that mining claims are property; and yet the Legislature declares that aliens shall not have " the same rights in regard to 'their' possession and enjoyment" as native-born citizens. The foreign miner act begins thus: " No person, not being a citizen of the United States, or who shall not have declared his intention to become such, prior to the passage of this act, (California Indians alone excepted) shall be allowed to take gold from the mines of this State, unless he shall have a license therefor." The Supreme Court have decided that the foreign miner license is constitutional, and their decision settles the law, but it does not make the right or convince the reason. The decision asserting the constitutionality of the act against foreign miners was rendered some years ago, and I think the present Judges might reverse the judgment of their predecessors.

If, however, the foreign miner act be constitutional and valid, it is clear that under it, every alien who pays his license is entitled to take up and hold mining claims, in the same manner and to the same extent as citizens. That right is promised when the license is exacted, and the Government is in honor bound to protect aliens in the exercise of that right so long as the Constitution and laws remain as at present. The mining regulations of many of the mineral districts prohibit the holding or working of claims by Chinamen; but this prohibition must be held void by the Courts, though it may be enforced by the power of the mob, against which the officers of the law are often powerless. The law may authorize a Chinaman to hold and work a claim, but if all the miners of a district tell him he shall not hold or work it, prudence will require him to obey their orders. These mining regulations have not been subjected to much trial in our Courts, and we have no authoritative guide as yet about their legal force in such cases.

Foreigners mining on private lands, and holding a lease from the land owner, are not subject to the foreign miners' tax.

Authority of Miners' Regulations. ₰ 159. In the case of the *Dutch Flat Water Company vs. Mooney* and others, (12 *Cal.*, 534) the Supreme Court of the State said:— "We do not decide the question as to the power of a mining district to pass a valid regulation, declaring the tenure of this species of property [mining claims] to be different from that created by general law." Since that decision, nothing has been done touching the matter by either the Legislature or the Supreme Court, and, therefore, the question still remains in doubt. I presume, however, that the mining regulations cannot set aside the statutes of the State.

In a previous case (*Waring vs. Crow*, 11 *Cal.* 366) the Court had said: "The plaintiff's right having been fixed by these rules of property, which are a part of the general law of the land, could not be divested by any mere neighborhood custom or regulation."

The foreign miner act does not extend to Utah, nor is there any probability that any similar statute will ever be adopted there, where wealthy capitalists and not roving placer diggers will control the legislation relating to the mines. Many aliens are now among the owners of the richest lodes in the silver districts.

Mode of taking up Claims. § 160. Any man entitled to mine may take up a claim for himself or for a company. If for himself, he goes to the Recorder, gives him a description of the claim, and pays the fee. The Recorder copies the description of the claim, which thereby becomes the property of the claimant, subject to the mining laws.

If the miner wishes to locate claims for a company, he gives a list of the names of the members of the company to the Recorder, and describes a tract of ground as large as that number of persons may be entitled to hold. If there be nine members in the company, and each man may hold one hundred feet square, then the claim may be three hundred feet square or nine hundred feet long by one hundred wide. In auriferous quartz or silver districts, the claims usually have a certain length on the lode, with as much ground on each side as may be necessary for working. The ordinary length of a claim for one man, in the silver districts, is two hundred feet. Sometimes two parallel lodes are found within ten feet of each other, and in such cases great inconveniences may ensue; for one company has no right to claim both lodes, and yet the two lodes are so near together that the works and workmen of one lode may be in the way of those of the other. In some quartz districts of California, a claimant not only gets the lode, but two hundred and fifty feet on each side of it, with all the minerals and veins. This secures him in the possession of abundant elbow room for his working. When the locator of a claim for a company makes his appearance with a list of names, the Recorder puts them down without inquiry. He has no authority to inquire where the persons live, or whether they live at all. The fact that they are not miners or have never been in the district, or are women or children, even if established, does not prevent the claim from being good and the title perfect, if made in accordance with the mining laws. It is not necessary to render a title perfect that the owner should ever see it in person. In some districts the miners' regulations require the Recorder to visit the ground before he enters a claim on his books. This is a wise provision and ought to find a place in every code. It is the only certain method of preventing a conflict of claims, which otherwise is likely to occur.

Incorporated Mining Companies. § 161. Many mines, including most, if not all the important silver lodes in Washoe, are wrought by joint stock companies, incorporated under the laws of California. The causes why they are incorporated in California, are, that most of the owners live in California; that most of the capital for opening the mines goes from here; that there is no law of incorporation in Utah, suitable to the case; and that Salt Lake City is far more remote in distance, and in business communication, from Washoe, than is San Francisco. These joint stock companies are formed in the following manner: Some persons—not less than three in number—owning most, or all of the feet in a claim, meet and agree to form a company. If there be seven members in the company, they may own 1,400 feet of a lode. They may value the lode at one hundred dollars, or five hundred dollars a foot, always putting the estimate high enough. Usually they make a share for each longitudinal foot, and multiplying their estimate of the value of a foot by the number of feet, they get the amount of the capital stock. Thus, 1,400 feet, would make 1,400 shares, and at five hundred dollars a share, the total capital stock of the company would be $700,000. The formation of such a company, and its incorporation under the laws of California, does not prove that the lode is worth a dollar, or that the company own a dollar, individually or collectively. When they desire to be incorporated, they draw up a certificate, which states the name of the company, its purpose, the time for which it is to exist, the amount of its capital stock; the number of shares, the value of each share; the names of the first Board of Trustees; the term for which they are to serve; and their principal place of business. This certificate, signed by about half a dozen persons, of those forming the company, (the signatures usually including all the Trustees) acknowledged before a notary, or officer authorized to take acknowledgments of deeds, is filed in the office of the County Clerk, and an original duplicate is filed in the office of the Secretary of State. They are thus fully incorporated; can adopt a constitution and by-laws; elect officers; and go to work. Their first official act is, to receive a deed of trust of the whole lead, and then issue stock to those who made the deed, giving to each the number of shares of stock corresponding to the number of feet which he owns on the mine. The deed or deeds of trust should be made after the incorporation of the company; but may be made before, when the list of Trustees has been settled upon in advance. The corporation law of California says that the stock of companies incorporated for mining purposes, " shall be deemed personal estate, and shall be transferred in such manner as may be prescribed by the by-laws of the company; but no transfer shall be valid, except between the parties thereto, until the same shall have been so entered on the books of the company as to show the names of the parties by and to whom transferred, the number and designation of the shares, and the date of the transfer." Whether these principles will be adopted in Utah, is not yet finally established; and I think prudence requires that transfer of valuable interests in mines should be made under seal, acknowledged and recorded, as deeds of other real estate. These formalities should be observed in California, also, in the transfer of interests in lodes, which are owned by companies not incorporated. In claims held by companies, the members are said to own a certain number of feet; but the interest is undivided; and a man owning one hundred feet in a claim 1,400 feet long, really owns a fourteenth part of the claim; is entitled to a fourteenth of its profits, and is liable for a fourteenth of its debts; but he is not exclusive owner of any part of the claim. When he becomes exclusive owner of any piece of ground, the company ceases to exist, so far as he is sole proprietor.

Decisions of Supreme Court. § 162. Many principles of mining law have been determined by the decisions of the Supreme Court of California, and in the succeeding paragraphs I state some of these, referring to the cases in which the principles were set forth. As the abbreviations of reference may not be understood by some persons, I will explain. The reference to (" *Fitzgerald vs. Urton*, 5 *Cal.* 308 ") means that the principle was stated in the opinion rendered by the Court, in the case of Fitzgerald against Urton, a report of which will be found on the 308th page of the fifth volume of the official reports of the Supreme Court of California.

Mining Laws in Nevada. § 163. There are no statutes, or published decisions in Nevada, relating to mining; but Judge Cradlebaugh, the Federal Judge of the district including Washoe and part of Esmeralda, has decided that the laws and usages of California, in regard to mining, will be considered as authorities in his Court. The principle is a sound one, and will give satisfaction to those owning mining claims in Nevada.

Mining on Pre-emption Claims. § 164. While the miner has no right to enter upon private land, held in fee simple, he may enter upon preëmption claims, and search for minerals. Congress has excluded the mineral lands from the operation of the preëmption laws; but many persons, nevertheless, have gone into the mining districts, and laid claim there, to tracts of land for farming and grazing purposes. In case the land is covered with growing crops, fruit trees or buildings, the farmer may require the miner to give bond to pay for all damages done to the crop, trees or buildings. (*Acts of April*, 1852, *and April*, 1855.)

The miner, however, has no right to enter upon town lots, in actual occupation. (*Fitzgerald v. Urton*, 5 *Cal.*, 308.)

A miner has no right to take a ditch through a farmer's enclosure, at least, not near his house and corral. (*Burge v. Underwood*, 6 *Cal.*, 45.)

Limits of Claims Downwards. § 165. A placer claim goes down perpendicularly from the surface; and usually a quartz claim follows the lode as deep into the earth as it may go. On several occasions there have been disputes between placer miners and quartz miners, where the lodes of the latter have run under the claims of the former. But no question of this kind has ever come before the Supreme Court of the State; nor is any provision made for a determination of the title, in such a case, by any of the local mining regulations known to me. A suit involving this question came up before the District Court in Yuba county, several years ago. The Marysville *News* reported the case as follows:

"The case, which was tried before the District Court in Yuba county, without a jury, was that of *B. F. Reed et al., vs. Bruce Frye et al.* The plaintiffs claimed a quartz ledge in the ground of the defendants, under some rules made by an association of about ten persons, after the ground was taken up for placer mining by the defendants; the plaintiff claiming quartz mining to be a separate and distinct branch, and that a miner could not hold a placer claim and a quartz ledge at the same time, even though it be in his own ground, unless he claim it specially as a quartz claim, separately from his mining claim. The plaintiffs did not claim the mining ground in their complaint, nor establish any prior right to the ledge. The defendants proved their right to the ground, and that they knew of the existence of the ledge, and claimed it with their mining ground. They also proved that the law under which the ground was taken up allowed a miner to hold from bank to bank, in a ravine, and that no side claims were allowed to be taken up on a ravine in that district. His Honor decided that, when a claim was taken up for mining purposes, the occupant was entitled to all the mineral found in that claim, whether in the earth or quartz; and, also, that the owner of a quartz ledge could not follow it through the ground of another party, unless the ledge was located and its boundaries defined before the ground was claimed and legally located."

This suit, however, does not present the question fairly. Let us suppose that A locates a placer claim; and, subsequently, B locates a quartz claim on a lode which is found to run under A's claim and forms its bed-rock; who owns that part of the lode perpendicularly under A's placer claim? Or, let us suppose that a hundred perpendicular feet of granite lie between the bottom of the dirt in A's claim and the lode passing under it; who owns that part of the lode under A's claim? In the former case, I think A should be the owner; and in the latter B. These are cases, however, which will not often happen; for it is a general rule that where the quartz is rich, the placers are poor.

If, however, the quartz claim be located first, I presume there would be no doubt that the quartz miner would be entitled to the whole lode, without any restriction because of the subsequent location of a placer claim over a portion of the vein.

When mining claims are located, notice is usually given that they are taken for "mining purposes," and the words "placer" or " quartz lode " are not mentioned ; but the kind of ground indicates whether the claim is to go down perpendicularly to the bed-rock, or run at an inclination to the horizon with a vein.

While quartz claims ordinarily follow the lode, with its dips and angles, to the full extent of its depth, there are a few districts (including Grass Valley, I believe) where the quartz claims go down perpendicularly.

Fluming and Tailing Claims. § 166. There are claims which do not carry any right to the minerals. For instance : a company may claim a piece of land for fluming purposes, and a placer miner may subsequently lay a valid mining claim to the same land ; but he may not be able to work it until the flume is abandoned.

Another kind of a claim which does not give title to the minerals in the land, is a claim for a place on which to deposit tailings. Every miner owns the tailings which run from his mill or sluice, and he has a right to hold, as a tailing claim, as much land as may be necessary to contain his tailings; but another miner may make a subsequent and valid claim to the same land for mining purposes; but before being able to work the claim, he may have to wait until the owner of the tailings has washed them, for which a reasonable time, (it may be for years) in accordance with the customs of the district, will be allowed him. If a miner wishes to have a good title to a tailing claim, he must show that he intends to preserve them, and he ought to confine them within an inclosure and post up notices of his claim. (*Jones vs. Jackson*, 9 *Cal.*, 237 ; *O'Keefe vs. Cunningham*, 9 *Cal.*, 589.)

When a miner allows his tailings to run upon the claim of another, the latter becomes their owner. (*Jones vs. Jackson*, 9 *Cal.*, 237.)

Custom among the miners has given to every claim-holder the right to run off his tailings through the claim or claims below his ; provided that he do no actual damage. If any direct damage be done, he must pay for it. This principle is not stated in any statute, decision or mining code ; but it is custom. When land is held by fee simple title, the owner has the right to prohibit every man from coming on his land, or making any kind of use of it.

Conveyance of Mining Claims. ? 167. It is prudent in purchasing a mining claim to take a deed, and have it sealed, acknowledged and recorded, like an ordinary conveyance of real estate in fee simple. The Supreme Court appears to have held different doctrines, at different times, on this point. In January, 1857, the Court said : " Bills of sale not under seal," are insufficient to convey a perfect title to a mining claim. (*McCarron vs. O'Connell et al.*, 7 *Cal.*, 152.) In July, 1859, the Court said : " We are unable to see why, upon questions as to the occupancy of the public mineral land, a transfer of the right of the occupant to the possession—which is about all his claim to it—is not as good for all purposes, to the vendee taking possession, when evidenced by an agreement, as by a deed." (*Jackson vs. Feather River Water Co.*, 14 *Cal.*, 23.)

The title to a mining claim may be seized and sold under execution. (*McKeon vs. Bisbee*, 9 *Cal.*, 137.)

Water Subject to Claim. ? 168. The water, like the land, may be claimed for mining purposes; and general usage does not restrict a man to such small quantities, comparatively, as of the latter. Any one man, for instance, has by general usage the right to take possession of a whole stream and lead its waters to the mining ground, and charge his own price for it. He thus becomes the owner of the ground covered by his reservoirs, dams, ditches and flumes, and may maintain actions for trespasses upon them in the same manner as if he owned the land in fee simple. Ordinarily, however, these ditching enterprises are so expensive, that they can only be carried through by large companies. But there is no limit to the length of ditch or the extent of reservoir, or the amount of water which they may hold, legally. The miners' laws never place any limits on these points; indeed, the more extensive these works are made, the better it is for the districts which they supply.

The Supreme Court has said : " The ownership of water as a substantive and valuable property, distinct, sometimes, from the land through which it flows, has been recognized by our courts ; and this ownership, of course, draws to it all the legal remedies for its invasion. The right accrues from appropriation ; this appropriation is the intent to take, accompanied by some open, physical demonstration of the intent, and for some valuable use. We have held that there is no difference in respect to this use, or rather purpose, to which the water is to be applied; at least, that an appropriation for the uses of a mill stands on the same footing as an appropriation for the use of the mines. Each of these purposes, indeed, may be equally useful, or even necessary to the miners themselves. But the nature of the use may be important, as denoting the extent of the water appropriated. Water taken for a mill is not an article of merchandise, to be sold in the market; it is merely a motive power, and after it passes the mill and subserves its purposes, may be used as an aid to the working of the mines. But this last use must not be inconsistent with the prior right acquired by the mill-owner, so far as his necessary use is concerned. This right of water may be transferred like other property." (*McDonald and Blackburn vs. The Bear River and Auburn Water and Mining Company*, 13 *Cal.*, 220.)

Water is claimed, in the mineral districts, by ditching companies, which intend to sell the water to miners; by miners intending to use the water in working their own placer claims; by quartz miners, to drive their quartz mills; by owners of saw mills and grist mills, to drive their mills; and by farmers, for purposes of irrigation.

Every claimant has a right to all the water which he has converted to his use, and no other person has a right to take the water away from him. The first claimant has the right to lead the water whither he pleases. Nobody coming subsequently has a right to demand that it shall be led back into its natural bed. No subsequent claimant has a right to go higher up the stream and take away a portion of the water. (*Irwin vs. Phillips*, 5 *Cal.*, 140.)

After the water of a river or brook has been appropriated, a subsequent claimant may lead the water out of its bed above and use it for mining purposes, if he turns it all back again. (*Bear River Water and Mining Company vs. York Mining Company*, 8 *Cal.*, 327.)

When water from an artificial ditch is turned into a natural water course, and mingled with natural waters of the stream, for the purpose of conducting it to another point, to be there used, it is not thereby abandoned; but may be taken out and used by the party thus conducting it, so that he do not, in so doing, diminish the quantity of the natural waters of the stream, to the injury of those who have previously appropriated such natural waters. (*Butte Canal and Ditch Compaay vs. Waters*, 11 *Cal.*, 143.)

If a ditch be cut for drainage merely, another claimant may come, and by a public claim, make himself owner of the water, for the purpose of mining. (*Maeris vs. Bicknell*, 7 *Cal.*, 261.)

Older Claim has Preference. § 169. The holder of a claim has a right to work it without any hindrance from later claimants. Subsequent locators have no right to dam up the water so as to turn it back on a prior claim, nor to erect such obstructions as to deprive the older claimant of the fall which may be necessary to enable him to wash his dirt conveniently, and which he had when he first took up his claim. (*Sims vs. Smith*, 7 *Cal.*, 148.)

Work done in getting machinery for a claim, or making a drain to free it of water, is considered as work done on the claim itself. (*Packer et al. vs. Heaton*, 9 *Cal.*, 568.)

Abandonment of Water. § 170. If a miner, who has appropriated the waters of a brook, abandons its waters and allows them to flow into another brook, which has been claimed by others, these become entitled to all the water. (*Butte Canal and Ditch Company vs. Vaughan*, 11 *Cal.*, 143.)

Claims entitled to Water. § 171. If a miner has taken a claim on the bar of a river, or in the bed of a brook—that is, a claim of such a character that the water of the river or brook must be used to work it—then no person, coming after him, has the right to lead the water away, so as to deprive him of its use.

Claim of a Route for a Ditch. § 172. The survey of the line of a ditch, the planting of stakes along the line, giving public notice of intent to make a ditch, and the commencement and vigorous prosecution of the work, are sufficient to give possession of the ground to a ditch company, and to make them its owners, as against everybody but the United States. (*Conger vs. Weaver*, 6 *Cal.*, 548.)

Miners' Regulations. § 173. The local mining regulations are recognized by our statutes as valid. The act to regulate proceedings in civil cases says that "in actions respecting mining claims, proof shall be admitted of the customs, usages or regulations established and in force at the bar or diggings embracing such claim; and such customs, usages or regulations, when not in conflict with the Constitution and laws of the State, shall govern the decision of the action."

The " customs, usages or regulations " here mentioned, are found written out and formally adopted by the miners of every mining district in the State. The size of the district is very irregular, and is determined by the miners themselves. A district may be a mile square, six miles square, or it may have any irregular shape; and the shape is usually very irregular, including a certain class of diggings. The miners of this district call a public meeting and pass a code of regulations, describing therein the size of the claims; how many claims a man may hold; how the claims shall be recorded; what the recorder shall receive for each entry; how claims may be transferred; what the recorder shall receive for noting each transfer in his books; how claims may be forfeited; how much work must be done on a claim to maintain the title perfect; who shall be recorder; how disputes about mining claims shall be settled, and so on. Nearly every district has its recorder, who keeps a record of all the claims, and keeps a copy of the laws of the district open to inspection.

The laws in different districts vary in diggings of the same class; but the variance is still greater when a comparison of the laws of a quartz district is made with the laws of a tunneling district; or those of either with the code of a shallow ravine district, or the district of a flumed river.

The mining regulations of the silver districts bear much resemblance to those of the quartz districts.

There are certain general principles recognized in most of the mining codes. Among these are the following:

The discoverer of a district, or of a new class of diggings in it, is entitled to hold twice as much as ordinary miners. The principle might be said to be established as the general usage of the State.

Neglect to work a claim at a season when it is workable, is considered an abandonment and forfeiture of it, so that it may be seized by the first comer. In some districts, a claim must not be left more than three days; in others, not more than a week.

Stakes must be driven at every corner of a claim, and a paper giving a description of the boundaries of the claim, and the names of the owners, must be posted up on the claim.

In placer diggings, where the claims can only be worked during part of the year, when the water is either high or low, claims are not forfeited for want of work during the unpropitious season. A river claim may be left untouched during the winter, and a dry ravine claim during the summer, without danger to the title. It is not uncommon for the laws to provide that a man may hold one of each class.

Neglect to work on account of sickness of the owner causes no forfeiture.

In cases where expensive tunnels are cut or shafts dug, in ground where much doubt is entertained about the finding of paying diggings, the prospectors are sometimes authorized to hold two claims each.

A great diversity of custom prevails about the number of claims which a man may hold. One may be held by location; and in some places, any number by purchase.

Companies are entitled to hold as much as the individual members could hold if they were separate.

Work done on any part of a company's claim, secures the title to the whole of it.

The amount of work necessary to be done on claims to secure the title differs greatly; it may be a day's work per week, or two days' per week, in the working season.

In some districts it is necessary to dig a trench along the boundary of every claim, and to put up on it a tin sign marked with the number which the claim has in the recorder's book.

I give, in the latter part of this chapter, copies of the miners' codes in several of the districts of California and Nevada. Among these codes is that of the Genoa District, in Nevada. In this code are singular clauses, not found in any other mining regulations on the coast. The first peculiar and important feature of this code is that by the expenditure of $1,000 in working a claim, the claimant gets a title beyond all forfeiture. Having spent that amount, he is secure forever in his possessions, against all other miners; but, of course, the code cannot give title as against the Federal Government, which is the fee simple owner of the land and minerals. In every other code of mining regulations known to me, the title of mining claims is forfeited whenever the claim is abandoned; and neglect to work for a few days in the season when the claim can be worked, is considered abandonment. If it be politic to give a title that cannot be forfeited by neglect to work, it would be well to specify how the fact that $1,000 has been spent on a claim is to be ascertained, and how the title to a claim may be given up. The expenditure of $500 secures the claims against forfeiture for a year.

This code further entitles any man to hold three hundred feet on any vein in the district; and to hold two hundred feet of land on each side of his vein, so that he has an abundance of room for working. If, however, the lead run lengthwise with a hill, and the claimant cuts a tunnel into the hill, he owns all the ground through which his tunnel goes, no matter how wide the hill may be. Thus, if the hill be one thousand feet wide, and a tunnel be run entirely through it, at a cost of $1,000, he gets a perfect title to a claim three hundred by one thousand feet in size.

This code further provides that it shall not be altered except by a two-thirds vote of all persons owning claims in the district, and after twenty days' notice published in several newspapers. These are formalities not ordinarily required in such cases: but their object—to secure the titles of claims—is very plain.

Regulations of Columbia District. ¿ 174. The following code of mining regulations, adopted by the miners in the Columbia district, Tuolumne county, and revised by them in October, 1860, is considered one of the best and most complete in the State:

" ARTICLE 1. The Columbia Mining District shall hereafter be considered to contain all the territory embraced within the following bounds: Beginning at the site of M'Kenny's old store, on Springfield Flat, and running in a direct line to a spring on a gulch known as Spring Gulch—said gulch running in a southern direction from Santiago Hill. Thence, in a direct line from said spring, to the angle of the road leading from Saw Mill Flat to Kelly's Ranch, near Wood's Creek. Thence, running along the ridge on the west of Wood's Creek, to the southern bounds of Yankee Hill District. Thence, following the ridge, to the high flume between Columbia and Yankee Hill. Thence, following the New Water Company's ditch to Summit Pass. Thence, in a direct line to the head of Experimental Gulch—including said gulch. Thence, following the upland to a point opposite Pine Log Crossing. Thence, following the upland to the head of Fox Gulch, and including said gulch. Thence following the upland around the head of Dead Man's Gulch, to the site of the Lawnsdale Saw Mill. Thence, in a direct line to the place of beginning.

"Art. 2. A full claim for mining purposes, on the flats or hills in this district, shall consist of an area equal to that of one hundred feet square. A full claim on ravines shall consist of one hundred feet running on the ravine, and of a width at the discretion of the claimant; provided it does not exceed one hundred feet.

"Art. 3. No person or persons shall be allowed to hold more than one full claim, within the bounds of this district, by location; nor shall it consist of more than two parcels of ground, the sum of the area of which shall not exceed one full claim; provided nothing in this article shall be so construed as to prevent miners from associating in companies to carry on mining operations—such companies holding no more than one claim to each member.

"Art. 4. A claim may be held for five days after water can be procured at the usual rates, by distinctly marking its bounds by ditches, or by the erection of good and sufficient stakes at each corner, with a notice at each end of the claim, followed by the names of the claimants, and by recording the same according to the provisions of Article 10th.

"Art. 5. When a party has already commenced operations upon a claim, and is obliged to discontinue for want of water, or by sickness or unavoidable accident, the presence upon the ground of the tom and sluices, or such machines as are employed in working the claim, shall be considered as sufficient evidence that the ground is not abandoned, and shall serve instead of other notice; the bounds of the claim still being defined, except so far as the marks may have been obliterated by the work which has been done, or by other causes.

"Art. 6. Claims shall be forfeited when parties holding them have neglected to fulfill the requirements of the preceding articles; or have neglected working them for five days after water can be procured at the usual rates, unless prevented by sickness or unavoidable accident, or unless the miners have provided by law to the contrary.

"Art. 7. Earth thrown up for the purpose of washing shall not be held distinct from the claim from which it was taken; but shall constitute part and parcel of such claim.

"Art. 8. Water flowing naturally through gold-bearing

ravines, shall not be diverted from its natural course without the consent of parties working on such ravines; and when so diverted, it shall be held subject to a requisition of the party interested.

"ART. 9. No Asiatics shall be allowed to mine in this district.

"ART. 10. Any or all claims now located, or that may be located and worked, can be laid over at any time, for any length of time not to exceed six months, by the person or persons holding the same appearing before the recorder of the district, with two or more disinterested miners, who shall certify over their own signatures that the said claim or claims cannot be worked to advantage, and by having the same recorded according to the laws of the district, and by paying a fee of one dollar; provided, each claimant shall sign the record in person or by a legal representative, stating at the same time that said claim is held by location or by purchase.

"ART. 11. There shall be a recorder elected, who shall hold the office for one year from the date of his election, or until his successor be elected; whose duty it shall be to keep a record of all miners' meetings held in the district; to record all claims, when requested by the claimants, in a book to be kept for that purpose, according to Article 10th; and to call miners' meetings, by posting notices throughout the district, when fifteen or more miners of the district shall present him with a petition stating the object of the meeting, and paying for printing notices; provided, that in the absence of the recorder, the above named number of miners shall not be disqualified to call a meeting, at the place specified in Article 13th. He shall at all proper times keep his record book open for inspection.

"ART. 12. No company or companies of miners, who may occupy the natural channel through any gulch or ravine for a tail-race or flume, shall have the exclusive right of such channel, to the exclusion of any company of miners who may wish to run their tailings into the same.

"ART. 13. Any party or parties locating claims in gulches or ravines where such flumes or tail-races exist, shall first confer with the party or parties owning said tail-races or flumes, for the use of the same on such conditions as they may agree upon; and in case of a disagreement, each party shall choose two disinterested miners, and the four shall choose a fifth, who may determine the matter or matters in dispute.

"Art. 14. Any company or companies of miners shall have the right to run their water and tailings across the claim or claims below them, if it can be done without injury to the lower claims.

"Art. 15. The limits of this district shall not be changed without the consent of a regularly called mass meeting of the miners of the district.

"Art. 16. No miners' meetings held outside of Columbia, for the purpose of making laws to govern any portion of the district, or to amend these laws in any manner, shall be considered as legal.

"Art. 18. All mining laws of this district, made previous to the foregoing, are hereby repealed."

Regulations of Pilot Hill. § 175. The mining code of Pilot Hill, Calaveras county, is as follows:

"Section 1. Each tunneling and shafting claim shall consist of one hundred feet in width to the man, and running through the hill on a parallel line with the commencement of the tunnel.

"Sec. 2. That each company holding tunnel or shafting claims, in order to hold the same, shall be required to perform work to the amount of twenty-five dollars each week for a period not to exceed twelve months.

"Sec. 3. That each gulch claim shall consist of one hundred and fifty feet in length, by fifty in width, to each man.

"Sec. 4. That each surface claim shall consist of two hundred feet in length, by one hundred feet in width, to the man.

"Sec. 5. That each gulch and surface claim shall be worked within three days after the date of location, if water can be obtained.

"Sec. 6. That each tunneling, shafting, gulch and surface claim shall be marked off by stakes, or other marks, so that the boundaries of each claim can be distinctly traced."

Regulations of Mush Flat. § 176. The code of Mush Flat, (Placer county) is very long. The following abstract gives all its points:

Article first gives the boundaries of the district. Article second provides for the election of a recorder to record claims. Article 3d. If recorder wishes to vacate his office, must give miners of the district a week's notice of a meeting for new election. Arts. 4th and 5th. Hill, ravine, flat and side-hill claims, to be two hundred by three hundred feet in size. Evidence of location—stakes and notice. Art. 6th. A "preëmpted claim," to mean one in immediate vicinity of paying diggings. And not lawful for any one to hold more than one each of hill, flat, or ravine claims by preëmption. Art. 7th. Any one empowered to hold as many claims by purchase as desired. Art. 8th. A "prospect claim," to mean one disconnected from previously discovered paying diggings, and located for prospecting purposes. Art. 9th. One month's labor, or its value, expended in prospecting a claim shall admit of three additional months suspensions of labor without forfeiture of claim. Art. 10th. The recorder, and a person appointed by a prospector, shall be a committee to decide upon the amount and value of labor performed, as required by preceding article, and the claim and its boundaries duly recorded, along with affidavit of committee. The duplicate of this furnished prospector, and posted upon the claim. Art. 11th. At expiration of three months, prospector to perform three months of labor, which shall entitle him to a suspension of work upon claim for six months, by proceeding as required by Article 10th. Art. 12th. When unremunerative labor to the amount of $1,000 has been expended upon claim, prospector may discontinue operations one year, by following provisions of Article 10th. Art. 13th. One person may have several "prospect claims," if in different locations; but his individual labor to be employed only upon one. Art. 14th. A "prospect claim" not to be laid by, and another to be taken, without a final abandonment of the first. Art. 15th. "Prospect claims" to be worked within twelve days of location, and every twelfth consecutive day thereafter, until the amount of labor required by Articles 9th, 11th and 13th are performed, without sickness or inclement weather prevents. Art. 16th. Preëmpted and purchased claims, if not paying ones, entitled to same privileges as "prospect claims." Committee of inspectors, to decide if they be entitled to the privileges. Art. 17th. Paying claims to be exempt from forfeiture, must be represented every twelfth consecutive day, if a sluice head of water is obtainable. Art. 18th. Holder of paying claim may delay work to the second twelfth consecutive day, by satisfying a board of referees of the necessity for so doing. Said board to have the reasons entered upon record of recorder. Art. 19th. Board of referees, if unable to agree, in cases as above, shall select a third person, whose decision shall be final. None of the board shall be interested in the claim. Art. 20th. Upon transfer of any claim, the transfer must be made upon the recorder's books within forty-eight hours, to avoid forfeiture. Art. 21st. No claim shall be forfeited when the district is not supplied with water for mining, except those accepting the privileges of "prospect claims." Art. 22d. Recorder to keep book with these laws recorded; to be open to the inspection of any one. Art. 23d. Fees of recorder, when acting on committee of inspectors, fifty cents; for recording mining claim, fifty cents; recording certificate of committee, fifty cents; for duplicate of same, fifty cents; service on board of referees, twenty-five cents; recording certified excuse of applicant, fifty cents; recording transfer and location of any claim, fifty cents. Art. 24th. When meetings of miners required, public notice of a week to be given, and stating object of meeting. Art. 25th. Meetings to be organized by the election of president and secretary, and the proceedings to be placed upon the records by recorder. None but the holders of claims in the district to vote at the meetings. Art. 26th. The foregoing laws may be amended by a meeting, properly called, by a two-thirds vote of a quorum present. A quorum to consist of twelve persons, and to be legal voters of the district.

Regulations of New Kanaka Camp. § 177. The following are the mining regulations of "New Kanaka Camp" district, in Tuolumne county, exclusive of the clause describing the bounds of the district, and some others in regard to duties and fees of the recorder:

"Art. 2. Creek claims shall be two hundred feet in length, and from bank to bank.

"Art. 3. Gulch or ravine claims shall be two hundred feet in length and fifty feet in width.

"Art. 4. All claims on bars or flats shall be two hundred feet in length and fifty feet in width.

"Art. 5. It shall be required that all claims be worked one full day in three, when permanent water can be had, except in cases of sickness or legal cause.

"Art. 6. All miners are entitled to one claim by preëmption and one by purchase; provided, such claims purchased shall be, on investigation, found to have been obtained in a legal or *bona fide* manner.

"Art. 7. Chinamen shall not be allowed to own claims in this district, either by purchase or preëmption.

"Art. 8. All persons who find it necessary to cut a tail-race to their claims, shall have the privilege of cutting through any ground below them, owned by other parties, provided it will not result to the injury of such parties.

"Art. 9. It shall be required of all persons owning claims in this district, to designate the boundaries of said claims by digging a trench around the same.

"Art. 10. All disputes arising in regard to mining shall be left to arbitration, each party to choose one man, and in case of disagreement, they to choose an umpire.

"Art. 11. Arbitrators in all cases, for services, shall be paid for all time consumed, at the rate of three dollars per day.

"Art. 12. All claims may be laid over, by having the same recorded, from the time ditch water fails until it can be obtained again.

"Art. 13. A recorder shall be chosen, whose duty shall be to keep a book of records, with the number of each claim recorded, from one to an unlimited number. It shall also be the duty of said recorder to go on to each and every claim recorded, and post at either end of such claim a piece of tin, with the number stamped thereon, corresponding with the number on the book of record."

Quartz Regulations of Tuolumne County. § 178. The quartz miners of Tuolumne county adopted the following code of regulations in 1858, and they are still in force:

"Art. 1. The jurisdiction of the following laws shall extend over and govern all quartz mining property within Tuolumne county.

"Art. 2. Each proprietor or locator of a quartz claim shall be entitled to one hundred and fifty feet (150) in length of the vein, including all its dips and angles; also one hundred and fifty feet (150) on each side of said vein, together with the right of way on either side of said vein, to run tunnels and drifts any distance that may be necessary in order to work said vein. *Provided*, That the right to one hundred and fifty (150) feet herein granted on each side of the vein, shall not be deemed to conflict with or detract from the right of any subsequent locator, who may discover a vein *outside* of said one hundred and fifty (150) feet, from following *his vein* through said ground.

"Art. 3. The original discoverer of a vein shall be entitled to hold three hundred (300) feet in length on said vein, by virtue of discovery.

"Art. 4. No man shall, by virtue of preëmption, be entitled to hold more than one claim on the same vein, except as provided in Art. 3d.

"Art. 5. All quartz claims hereafter taken up or located, shall be plainly marked by notices posted, containing the claimants' names and the number of feet claimed.

"Art. 6. The parties locating a quartz claim shall put at least one full day's work on said vein in every thirty days, in order to hold the same. A day's work shall be eight hours' labor. *Provided*, however, that the sum of one hundred ($100) dollars expended on said claim, shall hold the same for six months from the date of its expenditure.

"Art. 7. Any individual, company or companies, erecting machinery for working quartz shall, by virtue of said machinery, hold the vein or veins belonging to said individual, company or companies.

"Art. 8. These laws shall be in full force and effect from and after the first day of September, A. D. 1858."

Silver Regulations of the Virginia District.

§ 179. The following are the laws of the Virginia Mining District, Carson county, Nevada territory:

"At a meeting of the miners of Virginia district, held at Virginia City, Sept. 14th, 1859, the following laws were adopted for the government of the miners of said district.

"Art. 1. All quartz claims hereafter located shall be two hundred feet on the lead, including all its dips and angles.

"Art. 2. All discoverers of new quartz veins shall be entitled to an additional claim for discovery.

"Art. 3. All claims shall be designated by stakes and notices at each corner.

"Art. 4. All quartz claims shall be worked to the amount of ten dollars or three days' work per month to each claim, and the owner can work to the amount of forty dollars as soon after the location of the claim as he may elect; which amount being worked, shall exempt him from working on said claim for six months thereafter.

"ART. 5. All quartz claims shall be designated and known by a name and in sections.

"ART. 6. All claims shall be properly recorded within ten days from the time of location.

"ART. 7. All claims recorded in the Gold Hill record and lying in Virginia district, shall be recorded free of charge in the record of Virginia district, upon the presentation of a certificate from the recorder of Gold Hill district, certifying that said claims have been duly recorded in said district; and said claims shall be recorded within thirty days after the passage of this article.

"ART. 9. Surface and hill claims shall be one hundred feet square, and be designated by stakes and notices at each corner.

"ART. 10. All ravine and gulch claims shall be one hundred feet in length, and in width extend from bank to bank, and be designated by a stake and notice at each end.

"ART. 11. All claims shall be worked within ten days after water can be had sufficient to work said claims.

"ART. 12. All ravine, gulch and surface claims shall be recorded within ten days after location.

"ART. 13. All claims not worked according to the laws of this district, shall be forfeited and subject to relocation.

"ART. 14. There shall be a recorder elected, to hold his office for the term of twelve months, who shall be entitled to the sum of fifty cents for each claim located and recorded.

"ART. 15. The recorder shall keep a book with all the laws of this district written therein, which shall at all times be subject to the inspection of the miners of said district; and he is furthermore required to post in two conspicuous places, a copy of the laws of said district.

"On motion, it was resolved that these laws be published in the *Territorial Enterprise* for one month."

Silver Regulations of the Genoa District.

§ 180. The mining regulations of the Genoa district, in Nevada Territory, are the following:

"SECTION 1. The district shall be known by the name of the Genoa Mining District, and shall be bounded as follows, to wit: On the north by the Eagle Valley District; on the east by the Carson river; on the south by an east and west line crossing at Mott's Mill; on the west by Lake Bigler.

"Sec. 2. The officers of this district shall consist of a President, Vice President, and Recorder, elected by those taking part in these proceedings, who shall hold their respective offices for the term of one year from the date of the election.

"Sec. 3. It shall be the duty of the President to call all necessary meetings of the miners of this district, to preside thereat, and to discharge the duties pertaining to said office.

"Sec. 4. In the absence of the President from the district, or his inability to act, it shall be the duty of the Vice President to act in his stead.

"Sec. 5. It shall be the duty of the Recorder to keep, in a suitable book or books, a full and truthful record of the proceedings of all public meetings; to place on record all claims brought to him for that purpose, when such claim shall not interfere with or affect the rights and interests of prior locators, recording the same in the order of their date; for which service he shall receive fifty cents for each claim recorded and forty cents per folio for recording the transfer, bill of sale or deed of and to any mining property. It shall also be the duty of the Recorder to keep his books open at all times to the inspection of the public. He shall have the power to appoint a deputy to act in his stead, for whose official acts he shall be held responsible.

"Sec. 6. All examinations of the record must be made in the full presence of the Recorder or his deputy; and in no instance shall any person or persons making examinations of such record be permitted to use pen and ink; the Recorder shall furnish to such person or persons examining the records a lead pencil, with which memoranda may be made.

"Sec. 7. Notice of a claim or location of mining ground, by any individual or by a company, on file in the Recorder's office, shall be deemed equivalent to a record of the same.

"Sec. 8. Each claimant shall be entitled to hold by location, two hundred feet on any lead in the district, with all its dips, angles and spurs, off-shoots, out-crops, depths, widths, variations, and all the minerals and other valuables therein contained; the discoverer of and locator on a new lead being entitled to one claim extra for discovery.

"Sec. 9. The locators of any lead, lode or ledge in the district, shall be entitled to hold, on each side of the lead, ledge or lode, located by him or them, two hundred feet. Any lateral veins, lodes or leads bearing minerals, within the space of the said two hundred feet on each side of the main ledge, shall be considered as claimed by and entirely belonging to the locator or locators of a lead and his or their assigns, and part and parcel of the same mine.

"Sec. 10. It shall be the privilege of any person, persons or company, when the vein, ledge or lode of mineral is not distinctly traceable upon the surface, to take up the ground they desire to prospect, stating in their notice the manner in which they intend to prospect the same, whether by running a tunnel, cut and drifts, or shafts, and the length or depth of such tunnel, cut or shaft and drifts. If the locator or locators who claim ground for mining purposes shall disclose their intention to run a tunnel into, through or across the line of their claim, then it shall be understood that the length of the tunnel which their notice declares it to be their intention of running, shall be and is hereby intended to grant and determine the width of claim located, and no person or persons shall be allowed to locate claims coming within the distance and bounds claimed by such notice of record.

"Sec. 11. Any person or persons, company or companies, who shall locate ground under these laws, shall be entitled to hold and enjoy and receive all the profit of working any and all leads, lodes or ledges of mineral deposits, found on and within the limits of their location and claim, by the running of tunnel, cut or shaft and drifts, on any part of the ground claimed, which shall be considered as work done upon the claims, and if such work amounts to that required by these laws, the title of the ground claimed shall be deemed to vest in the locators and their assigns.

"Sec. 12. Every claim, whether by individual or by company located, shall be recorded within ten days after the date of location.

"Sec. 13. Three days' labor shall be done on each claim, or on the company's ground for each claim, in order to hold the same, from this date until the first day of June, 1861. The work so required must be commenced within thirty days after date of record, and fully performed within ninety days after commencement.

"Sec. 14. Whenever work shall have been done upon the claims of a company, deemed to be of the value or cost of five hundred dollars or upwards, the claims on which such sums shall have been expended cannot and shall not be subject to a forfeiture and relocation for the term of one year from the date of the last work done.

"Sec. 15. Whenever one thousand dollars shall have been expended on the claims of a company in this district, the ground so claimed by the company shall be deemed as belonging in fee to the locators thereof, and their assigns; and the same shall not be subject to location or relocation by other parties ever after, except by an acknowledged abandonment by the company of the ground.

"Sec. 16. The Recorder shall go upon the ground with any and all parties desiring to locate claims, and shall be entitled to receive for such service, say for a company locating of ten or more names, five dollars; and for each location numbering less than ten names, fifty cents each.

"Sec. 17. Location of claims to mining ground may be made within this district, between the dates of October, 1860, and the first of May, A. D. 1861, and any such claims so located and recorded, shall not be subject to forfeit or deemed forfeited and subject to relocation, for not working within these dates. Claims located as in this section of the laws, must be represented, and work commenced on the first day of May next, or such claims shall be deemed forfeited and subject to relocation.

"Sec. 18. These laws contemplate and provide for the location of two classes of claims—one to be denominated 'croppings' claims, being ledges or lodes visible; and the other to be called 'tunnel' claims, indicating that the ledge or lode claimed is not distinctly traceable on the surface.

"Sec. 19. An election of officers shall be held, at a place within this district to be designated by public notice of the President, in one year from the date of the adoption of these rules and laws. A vacancy can be filled by an election, on the call of the President at any time, by giving ten days' notice. All officers to hold until their successors are duly qualified. No person shall be allowed to vote after the adoption of these laws and the election of officers, at any subsequent election, except such voter be an actual owner of a claim or claims in this district.

"Sec. 20. This constitution, rules or mining laws, may be altered or amended by a two-thirds vote of those owning claims in the district, at any time twenty days' notice of such intention shall have been given in the *Territorial Enterprise* and Sacramento *Union*, and shall have been posted in three public places in the district."

CHAPTER XII.

MISCELLANY.

Value of Gold According to Fineness. § 181. Pure gold, 1000 fine, is worth $20.67 per ounce; gold 500 fine, is worth $10.33; 600, $12.40; 700, $14.47; 800, $16.53; 900, $18.60. The gold dust, as obtained from the placers, is usually sold by the miners to the merchants and bankers of the town where they do their trading. The buyers of dust know the value of the dust from the different districts, and are usually acquainted with the sellers; so they know at what price they can make a profit. The buyers send their dust to San Francisco by Wells, Fargo & Co.'s express, which has an office in every town of note between the Colorado and Fraser river. The charge for bringing the dust to San Francisco varies from one quarter of one per cent., to two and a quarter per cent. From Yreka, the price is two per cent.; from Shasta, one per cent.; from Nevada and Columbia, three-eighths of one per cent.; from Placerville, one-quarter of one per cent., and so forth.

Gold Dust Trade in California. § 182. At San Francisco, the dust is sent either to the Mint, or to a private assay office. Dust is never sent away now, except in small quantities, and as a curiosity; before shipment, it is nearly all converted into bars or coin. At the Mint, the gold is melted, assayed, refined and coined. The charge for melting, assaying and refining, is fourteen cents per ounce; and no deposits are received in less sums than one hundred dollars. The charge for coining is one half of one per cent. All gold deposited in the Mint must be melted, assayed, refined and coined; the deposit cannot be withdrawn after assay. The depositor must wait from four days to three weeks, before he receives the value of his deposit in coin. He may have a refined bar, if he please; (and making refined bars comes under the head of coining) but refined bars are seldom made now.

At private assay offices, the gold is simply melted, made into a bar, assayed, and the value stamped on the bar. The charge for this is one-fourth of one per cent., if the sum be eight hundred dollars or more; two dollars, if less. San Francisco has, at present, no private refinery for gold; the last one was closed in November, 1860. The charge for melting, assaying and refining, was ten cents per ounce. At private assay offices, the owner of dust gets the money for it within twenty-four hours.

How Gold is Coined. § 183. The following is a description of the processes through which the gold goes in the Mint.

The first room in the regular order of the business of the Mint, is the deposit room. Here the metal is taken and weighed, and a receipt given. The gold is then taken to the melting room, where each deposit is melted separately, in a black-lead crucible, and upon the melted mass saltpetre and soda are thrown and stirred round to oxydize the base metals, and the gold and more sterling metals, thoroughly mixed, are cast into a bar. After being taken to the Weigh Room and weighed, it is ready for the Assay Department. The assayer, with a chisel, chips off a corner from the bar, and the chip is melted and cast into a button, to give a round form, so that it may be easily rolled out. It is rolled into a ribbon, and filed down till it weighs exactly ten grains, weighed by a scale which turns at the thousandth part of a grain. The ribbon is rolled up with sheet lead, placed in a cup called a cupel, made of calcined bone ashes, and placed in a heat sufficient to melt the gold, and the base metals, copper, tin, etc., are absorbed by the porous material of the cupel, or carried off in oxydation. The gold is then pure, except an admixture of silver. The button is again rolled out into a ribbon about as thick as ordinary letter paper, and boiled in nitric acid, which dissolves the silver and leaves the gold pure, which is weighed, and the amount which it has lost gives an exact measure of the quantity of impurity in the original bar. Thus, if the piece assayed weighs nine grains, then nine-tenths of the bar is pure gold; and the clerk in the Deposit Room can immediately give a certificate of the amount of coin due the depositor.

After the bars have been assayed, they are, as a general rule, thrown in together, indiscriminately, as the property of of the Mint. The first process in the granulating room, is to melt the gold with twice the weight of silver; and while melted, it is poured into water mixed with a little nitric acid, and the metal falls to the bottom in fine grains. The granulated gold is taken out, and cast into large stone or porcelain pots, holding about fifteen gallons of nitric acid. These pots sit in hot water, heated by steam, and the boiling acid soon leaves the gold pure from all silver, copper, lead, tin, zinc or other base metals. It is taken out, filtered, washed, dried, and again taken to the melting room, where it is melted with nine per cent. of copper, and one per cent. of silver, to make it of the standard alloy of nine hundred-thousandths fine.

The gold thus alloyed is run into bars a foot long, an inch thick, and of the proper width for coin, from an inch and a half for double eagles, down to half an inch for dollars. The bars are delivered over to the coiner.

The coiner's first process is to put the bars through the rolling mill, which has two heavy rollers of cast steel, ten inches long and eight inches in diameter, rolling together. The bars are thus rolled out a number of times, until they are nearly the proper thickness of the coin. The rolling mill is made so that the bars can be rolled out of any thickness. The bars, when rolled out several times, become somewhat brittle, and are then taken to the Annealing Room. This room contains a large furnace of brick-work, with long chambers to receive the bars, which are placed in copper tubes and heated to a cherry red. The gold is thus made softer and more ductile, and is again taken to the rolling mill, rolled sufficiently, and again annealed previous to being drawn. The bars cannot be rolled out to an exactly equal thickness, and to secure exactness in this respect, the bar is drawn through an orifice in a piece of steel, and this orifice being somewhat smaller than the bar as rolled, reduces the whole to the same exact width and thickness. The bar, not quite as thick as the coin, is taken thence to the cutting machine, which, by a punch, cuts out from the bar round pieces, a little larger than the coin. These pieces are called blanks. They are carried to the Annealing Room, again annealed and washed with soap and water. They are then taken to the Adjusting Room. Here each blank is weighed separately, and made the exact weight for the coin. If too heavy, the blank is filed down; if too light, it is thrown into a box to be re-melted. The work in this room is done entirely by women.

The adjusted blanks are run through the milling machine, which compresses the blank to the exact diameter of the coin and raises the edge. The purpose of making the edge thicker is to make the coin pile neatly, to protect the figures and to improve the general appearance. About two hundred and fifty blanks are milled in a minute.

The milled blanks are carried back to the annealing room, placed in an air-tight cast iron box and placed in the furnace to be annealed, so that they may take the impression well. When they are at a cherry red they are taken out and poured immediately into water with a little sulphuric acid. This softens and cleanses the gold. The blanks are taken out, washed with cold water, put into hot water again, taken out, mixed in with saw-dust, which is then sifted off and the blanks are dried and perfectly clean.

They are again taken to the coining and milling room, and stamped. The coining machine is elegant and massive. The blanks are placed in a tube or pipe, and from this the machine takes them one by one, puts them between the dies, stamps them, throws them out of the die and carries them down into a box, and they are then delivered to the Treasurer and are ready for circulation.

About one-fourth of the gold yield of the coast goes to the San Francisco mint, the only mint on the coast. The treasure exported from San Francisco in 1860, was $42,325,916; the gold deposits of the mint during the same period amounted to $11,219,209. About $31,000,000 were therefore exported in unrefined bars and in dust. A small portion of the export was silver. The number of deposits made at the mint in 1860 was 4,841, showing that the average value of the deposits was $2,271. The gold coinage of the year amounted to $11,178,000, of which $10,899,000 was in double eagles, and $179,000 in smaller pieces. Of silver, $264,000 was coined in half dollars, quarter dollars and dimes; and $216,678 in refined silver bars. No three cent pieces, copper or nickels have ever been coined in California, nor are they used in the country; and half dimes are very rare.

Private Coins of California. § 184. Previous to the year 1856, much gold coin was made in California by private persons. The coins were not made of the standard fineness required in the mint, but were usually about 880 fine, and contained no alloy save silver, which gave the coin a light yellow color. Previous to '53, there was such a lack of coin, that a large proportion of the private coin was in fifty dollar pieces called "slugs" or "adobes," octagonal in shape, about two inches and a quarter across and an eighth of an inch thick. These slugs, as well as all the private coin, have disappeared from circulation, and are almost as great curiosities in California as elsewhere.

Gold Export of California. § 185. The $42,325,916 exported in 1860, was sent to the following places:

New York	$35,661.500	New Orleans	$57,796
China	3,374,680	Sandwich Islands	40,680
England	2,672,936	Victoria, V. I.	25,100
Panama	300,819	Mexico	19,400
Japan	94,200	Costa Rica	3,145
Manila	75,660		
		Total	$42,325,916

The gold sent to New York is shipped on the mail steamers via Panama. The cost of the shipment is three per cent., including one and one-fourth per cent. for insurance.

The largest shippers are Wells, Fargo & Co's Express Company and Banking House, B. Davidson, banker and agent of the Rothschilds, and Alsop & Co., bankers. During 1860, W., F. & Co. shipped about $6,000,000, Davidson about $5,000,000, and Alsop & Co. about $5,000,000, these three houses shipping nearly two-fifths of the entire export of treasure. The treasure sent to China, Japan, Manila, the Sandwich Islands and Mexico goes in trading vessels.

Cheats in Mining. § 186. Mining is a precarious business, and should be undertaken by inexperienced persons with great caution. It is impossible to know the cash value of a metalliferous lead or a placer claim, and the occupation is one wherein people are peculiarly liable to be carried away by excitements and by the hope of making fortunes suddenly, to pay much more for claims than they are really worth. There are also many modes of deceiving buyers. Sellers of veins of auriferous or argentiferous quartz will procure specimens from richer veins, and represent them as coming from their own; or will select a few rich specimens of their ore and represent them as fair specimens of the whole, or when the rock is sent to the assayer, they will slip in some pure metal. Sometimes a rogue, to decrease the yield of auriferous quartz, will put grease into the battery, and placer miners not unfrequently "salt" their claims which they wish to sell, by putting in gold dust. If the claim contain stiff clay, it may be "salted" by shooting gold dust into it from a pistol. The Oroville *Record* tells the following as a statement of events that happened about December, 1860:

"A party not a thousand miles from Oroville had a quartz ledge and mill which it was found desirable to dispose of. Procuring some really valuable quartz, in which the precious metal was plainly visible, they announced their willingness for parties desiring to purchase, to test the ledge. This was accepted by 'party of the second part,' who proceeded to prospect quartz procured for the occasion, by the 'party of the first part aforesaid.' The party proposing to purchase knew the quartz was valuable on sight, but desiring to purchase cheap, were not particularly anxious to produce a rich prospect, and deposited a tallow candle or two in the arastra. The grease prevented amalgamation, and the rich quartz was duly crushed, then ground into impalpable powder, but produced a very diminutive prospect. The ledge was pronounced comparatively valueless, and was purchased cheap, by the prospecting "party of the second part, aforesaid." Of course, the purchasers found no more valuable quartz, and were broken in a few months."

John Arthur Phillips, in his *Records of Mining and Metallurgy*, (pp. 197–199) gives the following advice:

"The uncertain nature of metalliferous mining affords unusual facilities for making unscrupulous misrepresentations, and consequently, whenever, through the abundance of money or other favorable causes, the public mind becomes credulous, it admits the grossest misstatements without examination, and readily consents to pay exorbitant sums for properties which are probably altogether worthless, or at least of but little intrinsic value. Sooner or later, however, the truth is arrived at, and under the influence of a violent reaction, an industry that deserves well is denounced as a delusion, whilst the real circumstances producing the evil are either slurred over or forgotten.

"It is often asked if mining, on the whole, is a profitable

industry. It may be replied that it is not only profitable, but largely so, provided caution and judgment be exercised in selecting the mines, and due integrity, skill and economy displayed in their management. If, however, these conditions be not fulfilled, the most disastrous consequences may be anticipated; since worthless undertakings will in this case be supported by the public, and after subscribers have paid extravagant premiums for indifferent properties, the capital necessary to develope them will either be squandered or injudiciously spent. * * * A capitalist wishing to become associated with a company prosecuting a mining enterprise, should first inquire into the character of the district in which the mine is situated; secondly, the honesty and ability of the person reporting on it; and thirdly, ascertain if the shareholders generally are in a position to meet the demands which a vigorous trial would be likely to impose on them. The constitution of the undertaking should also be investigated. Free shares are generally objectionable, and the system of giving a large sum of money for an untried property is highly pernicious; since in addition to its effects upon the ultimate profits of the concern, it encourages misstatements, and tends to support a class of unscrupulous speculators, through whose agency this branch of industry has been chiefly brought into disrepute.

"All mineral explorations should be conducted as rapidly as possible, for the purpose of lessening the aggregate amount of dead charges, and a practical, intelligent and honest man should be entrusted with the direction of the works. It is by no means essential that such a person should possess an elaborate education; but his ideas relative to the exigencies of his profession should be clear and well defined, and he ought, moreover, to be familiar with the use of the dial and with all the various operations of dressing and preparing the ores for the market. Grave mistakes are frequently committed by entrusting the local management of mineral undertakings to men who possess but a superficial acquaintance with the subject, and who sink large sums of money in the multiplicity of their schemes for economizing expenditure. Such persons often make extensive surface erections before proceeding to the development of the underground works, and from mere love of display absorb an undue proportion of the capital, forgetting that their arrangements can only be valuable in proportion as the mine itself becomes productive. Hence the undertaking becomes prematurely embarrassed, and is sometimes obliged, from this very cause alone, to terminate its existence.

"The radical principle to be observed by the capitalist, when about to invest money in profitable or established mines, is to distribute his means over a considerable number of them; since to adventure in any single concern is, with the best advice, a matter of considerable risk. Moreover, it is not judicious to confine investments to one district or to one class of mines alone; but to select the richest localities, and according to their importance, give the best mines they contain a *pro rata* proportion of capital. It should also be borne in mind, that certain adventures are so worked as to afford regular periodical dividends, while others only yield returns at irregular intervals. These two classes of properties do not necessarily imply that different systems are pursued in their management. Permanent dividend mines are often more advanced in their explorations than those which yield irregular profits; and although the percentage of returns is steadier in the former, yet the probability of improvement in the market value may be considered greater in the latter; thus allowing capitalists to realize a profit in one case which could not be obtained in the other."

How the Miners Live. § 187. Most of the miners live in a rough manner. The proportion of those who work for wages varies from one-third to two-thirds. Not one-half of them lay up any money. Many earn money with ease, and spend it as fast as they make it. Men engaged in mining are not noted, as a class, for sobriety and economy. Their occupation seems to have an influence to make them spendthrifts, and fond of riotous living. Not more than one Californian miner in five has a wife and family with him. Most of the others are unmarried, and have no prospect of matrimony. The women are not in the State now to furnish them with wives, and before the women can come, the men will have grown old. The people as a mass, in the mining districts, are very intelligent. They came from all parts of the world, have seen much of life in multitudinous phases, and have profited by their experience.

Miners' wages are from two dollars and fifty cents to four dollars per day, without boarding; or from twenty-five to thirty dollars per month, with boarding.

Cost of Living. § 188. The cost of living in the mines of California is about twice as much as in the Eastern States. The prices of flour, sugar, coffee, tea, beef and ready-made clothing, are about the same at San Francisco as at New York; pork, fresh butter, eggs, fresh fruit, milk, chickens, and the labor of domestic servants, are about four times as dear. Boarding in the mining towns costs three or four times as much as it would cost in towns of the same size in the Eastern States. The prices of merchandise vary greatly in the mining towns, according to the supply and demand. Nearly all the imported merchandise consumed in the mines, is hauled in wagons from Sacramento and Stockton, at a cost of from one to three cents per pound, during the summer. In the winter, the roads to many districts are impassable for loaded teams, and if the supply should then run out, the prices rise to a high figure. Provisions and merchandise in Washoe and the mines of British Columbia and Washington Territory, cost from ten to twenty cents per pound more than in San Francisco.

Mineral Lands should be Sold. § 189. It has long been my opinion that the mineral lands, or a large portion of the land in the mineral districts of California, should be sold. Not more than one-fortieth of the land within the limits of the mineral region is now worked in mining claims, and nine-tenths of it never will pay to work for minerals. All this land is withheld from sale by order of the Federal Government, and the U. S. Surveyor has orders not to run a line of his surveys within five miles of any spot where miners are at work. It is considered a matter of the utmost importance in other countries and in other parts of the Union, that the people should own the soil. Ownership makes the people permanent, and induces men to get wives and comfortable homes; and permanence and the possession of families and homes make them temperate, economical, industrious and careful of their reputations. Without homes, families and permanent residence, they must be intemperate, idle, wasteful of their money, regardless of their reputations, and without hope of improvement in the future. This is unfortunately the condition of many of the miners of California at the present time, and this will continue to be the condition of a large part of the mining population, until they are tied down to the soil by ownership. The unoccupied land of the mineral districts might be sold without disturbing in the least any vested rights. There would be a great demand for the land; the buyers would become permanent settlers; absolute ownership in tracts of eighty and one hundred and sixty acres would induce the people to make permanent improvements; families would increase; the value of land would rise; and society and trade would acquire a stability similar to that of other states. It is true that much of the unoccupied land may be rich in gold, and some of the purchasers would become very wealthy; but their wealth would be anchored in the State, which would derive far more benefit than if the gold were taken out by a dozen men, one-half of whom should remove to a distant part of the world with their money, and the other half should spend their earnings in dissipation. A person who owns a large tract of land by fee simple title, could work it to far more profit than a multitude of others who would own it in small parcels, and who would have to work in haste. For this reason, I think that the sale of the unoccupied mineral land would increase the amount of the gold yield. But I doubt whether the interests of the State require that all the auriferous ground should be washed. In many places there are extensive placers which will pay four dollars per day to the miner, whereas a farmer could not make more than three dollars per day during the summer. But would it not be better that a farmer should have the land? The miner leaves an ugly heap of gravel, an eye-sore, a desolation, and worthless forever; whereas the farmer would make a beautiful home, growing more valuable every year, and he and the State would derive a profit from it for all time to come. The interests of the fee simple owner are thus the interests of the State, to which the vagrant miner is a natural enemy. The latter looks only to the present; the former to the future. This question is one to which justice has never yet been done. It is too comprehensive for me to argue it here; and I can only refer the reader, who desires to see something more on the subject, to a letter in the New York Daily *Tribune* of the seventeenth of March, 1858, or to an article in the *Hesperian* Magazine of San Francisco, for January, 1860.

APPENDIX.

The Divining Rod. I inadvertently omitted to mention in the proper place, under the head of "Prospecting," that in Washoe and California, as well as in many other mining countries, the divining rod is not unfrequently used as a means of finding metalliferous veins. The instrument used is a freshly cut fork of hazel, shaped like a V, with the arms about a foot long, and perhaps half an inch in thickness. This divining rod is held horizontally before the breast of the prospector, one arm in each hand, and the point in front. Holding it thus he walks along where he supposes there may be metallic veins beneath him, and when he is over a lode, the point of the divining rod turns down. He then tries other places near until he determines the course and position of the lode. Only a few persons can use the divining rod; for the great majority it resolutely refuses to move. The process is precisely similar to that of finding water by the divining rod, and it is used in all parts of the United States and Europe. By scientific men, and the more intelligent people generally, belief in the divining rod is considered superstitious; but in the mineral districts of Wisconsin, Iowa, Michigan and Nevada Territory, it commands the faith of many persons who are neither ignorant nor grossly credulous. I am not prepared to ridicule the use of the divining rod. I know many men much respected in San Francisco for honesty and good common sense, who declare that they have seen excellent evidence in Washoe of the trustworthiness of this mode of prospecting. Besides, a plausible explanation of the principle involved has been submitted to the world by Baron Reichenbach (see especially page 60 of his *Odic-Magnetic Letters*, translated by John S. Hittell) and until his theory of Od, which he has set forth with great clearness and advocates with remarkable learning and ability, be overthrown, it does not become anybody to ridicule the divining rod.

Potosi. The Marysville *Express* gives the following account of the Potosi mines:

"The new mines are situated in New Mexico, about fifteen miles north of the California line, and two hundred and eighty miles distant from Los Angeles, from which place all the supplies are obtained. The exact locality of these mines is latitude 35° 51' north, and longitude 115° 50' west from Greenwich. Potosi is situated near the Salt Lake and Los Angeles road, about five or six miles south-east from the Mountain Springs. There is a fine wagon road all the way from Los Angeles to Potosi, with the exception of Cajon Cañon, where there is a steep hill sloping up the western side of the mountain a distance of about one hundred yards. On the other side of the mountain there is a gradual and easy grade. As the road now stands, 4000 pounds of freight are hauled through Cajon Cañon by light six-mule teams. A good road with an easy grade can be made through this cañon, and we observe that preparations are making to improve it.

"Going from Los Angeles, there is plenty of grass and water till the road leaves the Mohave at Camp Lake, a distance of one hundred and fifty-two miles. About eighteen miles east of Camp Lake are Mule Springs, containing water until late in the season. About fifteen or eighteen miles further east are the Bitter Springs, near which mustang grass grows. Further east, a distance of about forty-five miles, are the Coyote and Kingston Springs. Thence to Potosi is about forty-five miles. The Colorado river is about thirty-five miles distant from Potosi.

"Freight by land has ranged as high as eleven cents per pound; but recently some has been hauled as low as eight cents.

"All the leads in the Silver Butte are the same class of minerals. Numerous assays have been made at different times and places. They run from $10 to $1500 per ton, and some others as high as $4000 per ton have been reported. The ore contains silver, gold and lead. Some assays indicate 1,500 pounds of lead to the ton, but this is much above the average. The average assays in silver are from $500 to $900 per ton. In the Silver Buttes the ore is encased in limestone rock, and is very easily worked.

"Timber for building purposes is scarce; but there is an abundance of wood for all practicable purposes in the immediate vicinity. Plenty of water can be obtained for mining purposes, though it is not abundant. There is a fine spring on each side of the mountain—one in Potosi in the shape of a well. It affords an abundance of limestone water. Potosi is the name of a town located in the midst of the mining settlement at the Silver Buttes. It is regularly laid out between two spurs of the mountains, and adjoining Silver Butte.

"Numerous other discoveries of silver, gold, copper, iron and tin have been made within forty to sixty miles of the Silver Buttes; and an extensive vein, very rich in antimony and lead, has been discovered about forty miles north, in the Charleston mining district.

"South of the Silver Buttes, about twenty-five miles, in the same range of mountains, some minerals have been discovered. About sixty miles in a south-westerly direction, in California, is the Salt Springs gold mines. They consist of vast quartz leads penetrating the mountain, and are said to be very rich. They were worked in 1852, but were subsequently abandoned for some cause; but they are now worked by a Los Angeles company, with Mexican labor, arastras being used to crush the quartz.

"A line of stages has been recently started between Los Angeles and Potosi. The proprietor, Mr. Johnson, intends to connect an express with Wells, Fargo & Co.'s at Los Angeles, and thus supply facilities for getting letters and papers in the mines. The stages will run two or three times a week each way.

"The facilities for getting to the mines are now very good. A person can go by stage or steamer from San Francisco to Los Angeles, and thence by stage to Potosi. The cost of the entire trip will be about seventy dollars."

Esmeralda, in March, 1861. A correspondent of the Grass Valley *National* wrote thus from Aurora, the chief town of the Esmeralda district, under date of March 31st, 1861:

"From the fact that our mining laws do not require more than two days' labor to be done on each claim, prior to the first of June, 1861, little or no work has been done in the way of prospecting or developing claims heretofore discovered. Yet there is a great deal of hen-scratching upon the surface of the ground for indications, and consequently there is a large number of claims located upon *boulders* and spurs, or on ledges that never had any existence, save in the imagination of the excited prospectors. This carelessness, on the part of miners, in prospecting and defining ledges, entails a very unnecessary outlay for recording, and much of this wild-cat stock is sold to innocent parties in California, at high rates, when in fact it is not worth the recording fees. I am informed by Mr. Brawley, the recorder of mining claims, that there are upwards of five hundred ledges located and recorded, and over 1600 companies, numbering some 13,000 claims. By the mining laws of the district, he is allowed one dollar per claim for recording; fifty cents for a certificate of claim; two dollars and fifty cents for recording deed. From these prices, I should judge he has already realized, in the short space of seven months, the handsome sum of $16,000.

"Four-fifths of the claims located in the district are on Silver, Middle, and Last Chance hills. The principal and most noted ledges, are the Esmeralda, Antelope, Old and Young Winnemucca, Wide West, Utah, La Platte, Lady Washington, Lady Dibble, Wyoming, Silver Age, Rio Del Monte, Venuango, Constitution, Lucerne, Last Chance, Yellow Jacket, Eldorado, Live Oak, General Jackson, Ne-plus-ultra, and Sacramento.

"Aurora is the chief town in the district, and is pleasantly situated on a beautiful flat, formed by the conjunction of two ravines, which come down from between Silver, Middle, and Last Chance hills. The town was surveyed in the early part of last October by Mr. Clayton. The streets run east and west, and north and south, forming right angles; have a convenient width, (sixty feet) giving it a decided advantage over most Californian mining towns, in case of fire. Good water, and plenty of wood, with rich diggings surrounding it, of gold and silver, must make Aurora one of the most wealthy and permanent towns on the eastern slope. The present population is about six hundred, (five ladies, only) and increasing daily. There are about two hundred houses and cabins ten provision stores, eighteen or twenty whisky mills, two tin shops, (one just started by a Grass Valley man) two shoemakers' shops, two barbers' shops, two drug stores, one notary public, two commissioners of deeds, four lawyers, three doctors, one post office, two express offices, three butchers' shops, one

9

record office of mining claims and deeds, one hotel, three boarding houses, and two gambling saloons, with about one hundred and fifty bummers and loafers of the first water, hanging about, waiting for something to turn up; or waiting for some new diggings, so as to rush in and get a claim without having to prospect for it.

"New buildings are going up on every side, and active steps are being taken to build some four large quartz mills. A part of the machinery is now, I learn, on the way to this place— also, three saw mills at the Meadows, and one at Mono.

"Lumber is now worth two hundred dollars per thousand feet; flour, plain, sixteen dollars; self-raising, seventeen dollars per hundred pounds; bacon, thirty-five cents; California eggs, one dollar per dozen; Salt Lake eggs, seventy-five cents; butter, seventy-five cents; beans, twenty cents; syrup, three dollars per gallon; potatoes, twelve cents per pound, by the sack; hay, by the bale, fifteen cents, retail twenty cents; freight, from Placerville here, is twenty to twenty-two and one-half cents per pound. Provisions are plenty at the above rates."

Diamonds in California. We understand that Mr. John S. Bassett, of Cherokee Ravine, found, a few days since, in his mining claim at that place, what has been pronounced a diamond by Mr. Young, a jeweler of this town. It was found by him while puddling clay for washing. We understand, also, that S. Glass, Esq., some time since found two stones of the same description, which he sent to New York, and which were also pronounced diamonds.—*Oroville Democrat*, April, 1861.

Silver By-Laws. The certificate of incorporation must be considered as the constitution or charter of a company, which must then have by-laws to provide for the details of government and management. The following is a copy of the by-laws of the "Sides Gold and Silver Mining Company," which has a silver claim in the Virginia district, Nevada Territory. These by-laws were adopted in February of this year, and are probably very similar to the by-laws of most other similar companies:

"In order to secure a perfect government of the Sides Gold

"*Resolved*, That the following rules and regulations be adopted; and we, the undersigned, pledge ourselves to abide by the same, and hereby authorize and impower the officers herein mentioned to execute and perform their duties as herein expressed, and to carry out the meaning and intention of the said rules and regulations.

"Section 1. This company shall be known as the "Sides Gold and Silver Mining Company," and its operations shall be confined to mining ground on the Comstock lead, or lode, in Virginia mining district, N. T.

"Sec. 2. A majority of stock represented shall constitute a quorum of any business, at any regular or special meeting.

"Sec. 3. Regular meetings shall be held on the first Saturday of every month, in Virginia City. Special meetings may be called at any time in Virginia City; *provided*, a majority of the shareholders request the president to do so, which officer shall give five days' notice previous to the said special meeting.

"Sec. 4. The officers of this company shall consist of a president, who shall act as secretary and treasurer for the company; also, three others, with the president, who shall constitute a board of directors, who shall hold their offices during the pleasure of the company.

"Sec. 5. The president shall have a general supervision of all work required to be performed by the company; he shall act as *ex officio* president, secretary and treasurer, and shall preside at all meetings held by the company; give notice of and collect all assessments levied by the company; keep a true and correct account of all moneys received and disbursed; and at the end of each quarter shall present the company with a correct statement of its affairs.

"Sec. 6. All assessments shall be levied by a majority of the stock represented, and shall be due and payable within ten days from the time of levy.

"Sec. 7. Should default be made in payment of assessments, the president shall have power to sell so much of the interest of the party in default, to any person or persons, at public vendue at the works of the company, to pay the assessment thereon, in cash, first giving ten days' notice of the time and place of sale, by a written notice at the company's works, and also at two other public places in Carson county, N. T. And said president is hereby empowered to execute a sufficient deed of conveyance to the purchaser thereof, setting out fully the facts of the case.

" Sec. 8. No conveyance of any share of this company shall be considered valid, unless the party or parties purchasing shall freely and voluntarily agree to be governed by these laws; and every new member so purchasing shall be required to affix his or her signature hereto.

" Sec. 9. Any or all of the foregoing laws may be altered, amended or annulled at any regular meeting, provided a majority of the shareholders deem it advisable.

" Sec. 10. The foregoing laws and regulations shall become null and void as soon as the Comstock lead or lode is 'struck,' or before, if the company shall deem proper."

The Management of Washoe Mining Companies. A correspondent of the San Francisco *Bulletin*, writing from Virginia City under date of April 8, 1861, says:

" I do not know of any calling so excitable as that of mining. Men from all countries meet promiscuously, upon common ground, in search of gold and silver; they have but one object in view. In this desire there are no secessionists—all agree and harmonize—they have no North, no South, no East, no West; all are one, and are hugely delighted with each other. Beyond this one idea, all lack system and economy. The work is frequently undertaken just for the chances, regardless of cost; plans are matured, companies organized, and the wherewith raised to prosecute the work to a successful result. We often hear of men succeeding wonderfully in amassing means in such wild, unmatured operations; but seldom, if at all, of those who fail. The few who do succeed are immediately heralded to the world by the press, made heroes of, and almost worshipped as if they were demi-gods, no matter how worthless or how much favored by luck; whilst the good, intelligent, hard-working and unfortunate ones are possibly unnoticed, unless, indeed, to be found fault with by friends near and relations far away. The reverse of this should be the case; the hard-toiling miner should be talked kindly of, and encouraged by every act and gesture to be of good cheer to the end.

" That I may be better understood, permit me to state the manner of forming mining companies here. It is somewhat after the California style. Some two or twenty persons—the larger number being very objectionable on account of the large size of the claim, as well as the difficulty of collecting assessments from many of its members, who are scattered all over

God's creation—get together and agree upon their plan of operations. They elect a superintendent, give out contracts for prospecting the ground, levy assessments of so much per foot on each share, and then collect said assessments, if it can be done. Seldom is this accomplished after the first two or three installments, and in consequence has to be stopped, and perhaps finally abandoned. Although by-laws are made and signed by the members, yet there lacks all legal power to compel delinquents to pay up their assessments. This is but one of the ill consequences of men entirely unknown to each other forming companies, and relying upon the good faith of each of its members to discharge his pledged promised duty.

"To correct this shameful abuse, we need a law giving, by short, easy and cheap process, the right to the officers of the company to go into court, and on proper showings prove their account and have their remedy thereafter in execution against the property of all delinquent shareholders. No defaulting member could in justice complain against this mode of procedure. In this particular, and in relation to mining companies, the common law partnership principle should be superseded; the wants and protection of the honest miner, as well as the interest of the country, demand it. No law should exist, if it be found to aid men to take advantage of their fair undertaking, as well as also to retard the development of claims and the general prosperity of the country. As matters now stand, the existing law or custom is a principle working irremediable injury to all. According to the common law one partner can sue another on an account settled, though it be of the partnership. Why not, then, give the right to bring the parties into court, without compelling the party aggrieved to go into dissolution of the partnership, and adjust their accounts, leaving the party suffering to his remedy as in common law on an account adjusted?

"Another great evil exists. Companies when formed in the manner just stated, often find their hands tied by their faithless officers, who generally are elected for a given time, with a power to levy and collect assessments, and to superintend work and develop the claim. But it happens often that if any rich developments be made, the entire energies of the officers appear to become prostrated; the work barely drags on, and as a consequence, discontent arises among the members, which in turn aids these very officers in their speculative designs, and frustrates the very object of the company—the successful working of the claim. The officers in the meantime grow very complacent, and apparently endeavor to conciliate the discordant opinions which had been brought about designedly, by themselves, and which are indirectly kept up by outside influences set quietly in motion by themselves. In the meantime the undertaking slackens, the members grow careless of its interest, and scarcely if ever go near the works; the good ore is designedly passed by; but care is taken by the officers to take samples of the ore to their offices, where they are placed out of sight and shown only to their immediate confidential friends. Next comes the tragedy of confidence betrayed; the title of the company to the property is made apparently defective by this outside influence—the work of these very officers to frighten the stockholders to sell at any offered price. This succeeds admirably; the magnetic message meanwhile speeds away to San Francisco, Sacramento and elsewhere—'Buy! buy quick!' The order comes 'Buy all you can.' The purchase is made at a figure low—say ten or fifteen dollars—and sold by these very trustworthy (?) officers of the company, at say—just double that amount.

"This state of things is common street talk, and can, I honestly believe, be substantiated to a great extent. There are men here engaged in this same business, who, when they came, were without means, but who now own very large interests in claims believed to be as good as the Ophir perhaps, if the property were worked to develop them. But this is not the idea; procrastination, ostensibly to freeze out the small fish, is the plain, palpable object."

[The correspondent proceeds to find fault with the superintendent of a certain company, who receives handsome pay for doing little or nothing; and continues:]

"It may be asked why does the company allow such a man to take the management? Such of the members as are here do not consent to it; they object, find fault, and talk aloud in the streets of the abuse. But they are in the minority. In the sale of the stock, the majority is held in San Francisco, and the owners send here their proxies to the officers in charge to elect whom they please; and these always shape things to suit their present and prospective positions in the company, with which the owners below seem satisfied for the time being, doubtless waiting until the whole stock is in the hands of the capitalists. The whole scheme finally ends in a monopoly by an absentee majority.

"Although I have referred to the doings of a particular company here, yet these things are common to many companies. It strikes me that it would be to the final benefit of large stockholders in California to act in concert with the small stockholders here, who are the actual discoverers as well as the bone and sinew of the whole operation, to the end that the common interest of all might be secured. Let the fact of absentee monopoly be once fairly understood—an alarm among the now despised and badly used small stockholders will be the result—and, as a consequence, vengeance on the property of the men persisting in monopoly will be sought, in a way little thought of. I don't pretend to say that such conduct would be right; I only state the fact.

"These small stockholders, too, are now becoming heartily tired of incorporating companies, for the reason that they end in a monopoly by absentee shareholders. So soon as the deed of trust is executed, they find that they have surrendered their all. Heavy assessments immediately follow, and if not paid according to the programme, their stock is sold; and rather than suffer this, they sell at any price offered them, there being really no choice left.

"Another objection to these companies is, the money raised by way of assessments is frequently squandered in the building of fine brick and other buildings for the officers of the company, whom the controlling absent stockholders seem first most anxious to provide with exorbitant salaries, irrespective of merit or capability to arrange the affairs of the company, before the mine is in thorough working, paying condition."

Lost Mines of Arizona. In the last century, one of the most notable mines of what is now Arizona was the one called the *Planchas de Plata*, the "planks of silver." Its exact position is unknown now, though the neighborhood in which it was found is plainly indicated by the old records and letters. Don Manuel Retes, now Captain of the port of Mazatlan, thus spoke of this mine, in an essay on the mineral resources of Northern Sonora:

"This mineral deposit, situated $31\frac{1}{2}°$ north, long. $111\frac{1}{2}°$ west of Greenwich, is described by the Jesuits as having been the last century, distant from four to five leagues from the mine of Arizona; about fifteen from the town of Tumacaicori, the nearest settlement; about twenty-five from the Presidio of Santa Cruz; nearly ninety from Ures, and about one hundred and thirty from Guaymas. The silver was discovered in sheets of different size, from which the name of Planchas de Plata, 'sheets of silver,' originated. They were found almost on the surface, perfectly pure, and without adhering to any foreign substance; in a flexible state, capable of receiving impressions, and only hardening on being exposed to the atmosphere. The region which produces the same is an earth of the color of, and very similar to, ashes, which extends in visible leads more or less wide, and in parts subdivided in veins, over all the hills and mountains adjoining the main deposit. Among the sheets extracted, two are worth mentioning—especially one which on account of its almost fabulous size, weighing one hundred and forty-nine arrobas, it was found necessary to employ the heat of four forges at the same time to reduce to a smaller bulk—the other weighed twenty-one arrobas, though according to other accounts it was much larger. The news of such immense lumps having been found, without the investment of much labor, could not fail to convoke a great number of people to that region, not only from the neighboring settlements, but also from the most distant provinces. The amount of silver extracted within a very short period, amounted to 400 arrobas, or five tons."

Another mine very rich in silver was the Arizona; the position of which is also lost. It was in search of this mine that Count Raousset de Boulbon made his celebrated expedition into Sonora, whither he went, at first, in good faith and with peaceable intentions, though after he had been defrauded and attacked, he turned filibuster. There are persons who are ready to assert that the exact position of the Arizona mine is known; but the best informed say it is not.

THE END.

www.ingramcontent.com/pod-product-compliance
Lightning Source LLC
Chambersburg PA
CBHW021835230426
43669CB00008B/975